WERE THEY ON THE MOON IN 1969?

YES THEY WERE!

ERWIN KOSTOMAI

Prime Seven Media
518 Landmann St.
Tomah City, WI 54660

Printed in the United States of America

FOREWORD

FOREWORD HERE IN your hands is a book on the flights of Apollo spaceships to the Moon with manned crew. It covers content that includes all the facts from the idea to the final flight, and the references about the later opinions of this feat and of all the doubts that have been and are still arising—doubts about the authenticity of this event. In order to present to the reader the integrity of the story, I tried to make the book as complex as possible with all the facts in one place. In 1969, I had the opportunity to watch the direct transmission of the first landing on the Moon. I had the opportunity to think about it, dream and enjoy it. Today, this topic separates people. Only a scientific and wise approach enables us a proper insight into this event. The book is written in such a way that it can be read by both experts and laypeople, and everyone can find some interesting facts and basis for thinking. In the book you also notice the great competitive spirit of that time, not competition for money, egoism, but the competition of people for people. I want the readers to enjoy reading this book and I will be glad if some new thoughts come to their mind. The desire of the book is to offer to an attentive reader a different view also on contemporary topics. These are the topics that burden us and for which we do not yet have an opinion. The world has become so complicated that it makes life difficult for us, not easier, and despite the expectations, this world has not yet fulfilled the hopes of man. The judgment about this event will be pronounced by history and it will pronounce even greater judgment about the bias of the modern world, and the dangerous egoism that surrounds us. We have many opportunities, so let us stop the egoism and give room again to the arrangement. Let us not disappoint our history, let us not disappoint ourselves. Erwin Kostomai

1

Introduction of the book about landing on the Moon

I F THERE STILL exists the slightest doubt about whether a man was on the Moon in 1969 or not, let it be. To doubt means to think but this is not the right answer. If we doubt each and every thing, it can happen that in the dustbin of history, both complete truth and complete lie are encountered. And this means that development wouldn't be possible anymore, there would be nothing to believe in, what is however incompatible with human nature that is always seeking answers to ever-new questions. But what is there, what is there behind, what will become of us?

It seems that we have the same problem as Galileo Galilei had when he could not explain that Earth is not the center of the universe. They did not believe him because they were too arrogant and confident in their own self, or they were only concerned about their own benefits. Today, we have the same problem. Therefore, we are at risk of losing our awareness of this enormous progress, the organization and commitment of scientists, engineers, politicians, astronauts and other visionaries who made possible the landing on the Moon with human crew. Because of the refusal and disregard for the facts of some of the doubters that occurred in 1969, we have lost the impetus that we may have to start it again.

In the contents of the following book, besides the general description of the events, I would like to put special emphasis on three chapters. I would like the reader to deepen into three contents, namely, the cause of doubt of the conspiracy theorists, the evidence that unambiguously proves that they were on the Moon and the moral of a story that tells us what could be learned from it.

In addition to this, you will find in the book all the chronology of the events since the beginning of the human launch of satellites into the earth's circumference, to the idea and the execution of the journey to the Moon with a human crew. Following a detailed reading of this chronology, we can take this as evidence, since there are the events stated for which everyone knows have happened and which confirm for themselves that they could not been able to pass all this enormous process without an epilogue. An explicit description of the events that followed was convincing enough, since there are moments of exceptional commitment of people, self-control, worker's zeal and precise engineering analysis described, which left nothing to chance in the creating of whole programme. These moments are supported by almost unbelievable data on the required number of people, the time and amount of material used to carry out the project. I will also deal with the explanations that have arisen in my mind with the use of healthy human sense. In a clear way, I will show what can be claimed or impossible to claim. This is the essence of this book, which can convince even the greatest doubters. Well, probably not all.

I will describe the probable causes of doubt. They are always caused by misunderstanding of things, because of sensationalism or other needs. I wrote the book in the name of the reason and progress of mankind that the achievements would not be ignored. By contempt, there is damage to all people on this planet, since better lives of all of us depend on such achievements. Better days for everyone will come when science is rewarded as it deserves. Not only those who are not scientists should benefit the most from state-of-the-art science.

I invite you to an interesting journey through the content, which will not leave anyone indifferent. I would like this book to enthuse us as mankind, what can be achieved with the combined forces, and what power we actually possess. And this power, as it seems, has almost no borders.

2

Chronology of the project from the beginning to the first landing on the Moon

2.1. Flights to the Moon from their beginnings until today

THE AMERICANS HAVE been on the Moon six times; on each mission there were three astronauts, and two of them directly landed and walked across the Moon, and the third was always circulating around it. There were 12 people on the ground of our natural satellite, since the Russians or other countries did not send people on the Moon, but only automatic spacecrafts.

The first successful launch of rockets was towards the end of the 1950s, followed by flights with ‹animal crew›. In April 1961, Yuri Gagarin became the first man in space and the first to circle the Earth. Since this period was marked by a competition between the United States and the Soviet Union, both of the superpowers (the USA even more due to the ‹defeat› Gagarin inflicted on them) aimed to send a man to the Moon. For this purpose, they have set up the Apollo project. In the first missions, Mercury, Gemini, they mainly tested different mechanisms to be used during the flight to the Moon. Apollo 7 was the first mission with human crew in this program. Apollo 8 was the first spacecraft with human crew to circle the Moon. The next big event was the landing of a man on the Moon. As part of the Apollo 11 mission, on July 20, 1969, Neil Armstrong was the first man who stepped on the Moon's surface saying: "That's one small step for a man, one giant leap for mankind." Armstrong was soon followed by Edwin Aldrin. After carrying out various experiments (and, of course, planting the American flag) and collecting soil samples, they returned to the lunar module. After a few hours, the lunar module docked with the command module in which Michael Collins was the third member of the mission and returned safely to Earth.

Picture 1 Edwin Aldrin stepping onto the Moon's surface

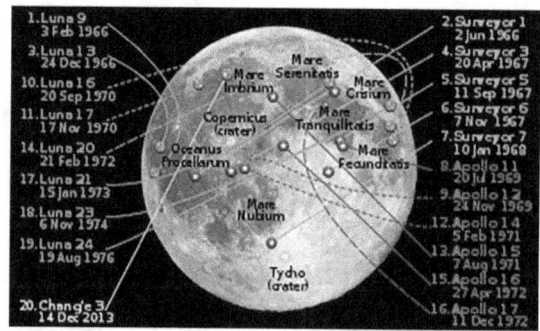

Picture 2 Landings of all expeditions to the Moon

In the following missions, they landed on the Moon five more times (until Apollo 17—with the exception of Apollo 13, where severe malfunction occurred). The table below shows some data for the Apollo 11 to 17 missions. Astronauts that have stepped onto the Moon's surface are marked with an asterisk.

Mission	Launch	Moon landing	Crew
11	16. July 1969	20. July 1969	Neil A. Armstrong Michael Collins Edwin E. "Buzz" Aldrin

12	14. November 1969	19. November 1969	Charles "Pete" Conrad, Jr Richard F. Gordon, Jr Alan L. Bean
13	11. April 1970		James A. Lovell, Jr John F. Swigert, Jr Fred W. Haise, Jr
14	31. January 1971	5. February 1971	Alan B. Shepard, Jr Stuart A. Roosa Edgar D. Mitchell
15	26. July 1971	30. July 1971	David R. Scott Alfred M. Worden James B. Irwin
16	16. April1972	20. April 1972	John W. Young Thomas K. Mattingly II Charles M. Duke, Jr
17	7. December 1972	11. December 1972	Eugene A. Cernan Ronald B. Evans Harrison H. "Jack" Schmitt

Table 1 List of the flights to the Moon with human crew

The last time a man set the foot on the Moon's surface was in December 1972. Although there are many plans to explore the Moon's surface, it is unlikely, due to enormous costs, that we will soon send the thirteenth man to the Moon.

Here is another interesting fact: In some books, we can read that they first landed on Moon on July 21 (not 20) 1969. The reason for this is the time zones on Earth. Armstrong made his first steps on the Moon on July 20, 1969 late in the evening Eastern American Time (22:56 PM EDT), while in Europe it was then already July 21st.

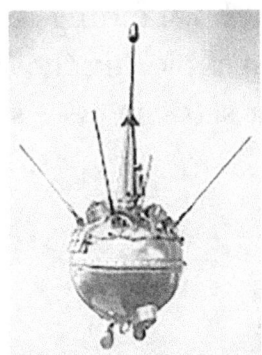

Picture 3 First satelite in Space

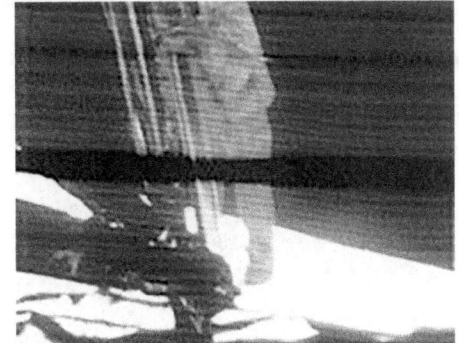

Picture 4 The first step of a man on the Moon

Space race has made progress from the launch of a small satellite to the first steps on the Moon.

Landing on the Moon means coming to the Moon's surface, the only Earth's natural satellite. Since the Moon is the closest celestial body to the Earth, and most accessible, it has been the logical goal of space programs from the beginning of the era of exploration of space.

Most of the expeditions ended with more or less controlled fall of robotic objects on the Moon's surface, with some probes crashed to pieces, and 20 expeditions made a soft landing. After landing, some of them carried out a launch of landing module from the Moon's surface and its return to Earth, six of which with human crew.

The first objects of human manufacture reached the Moon during the Cold War in the mid-20th century, when a frenzied space race took place between the then superpowers of the Soviet Union and the USA. On September 13, 1959, the Soviet probe Luna 2 became the object that crashed onto the Moon, and a few years later, the superpowers carried out, at an interval of several months, the first soft landing of robotic spacecrafts (the Soviet Union Luna 99 on 31 January 1966 and the US – Surveyor 1 on 30 May 1966). The peak of the race was reached by the American Apollo program with the first human landing on the Moon. It was succeeded by the Apollo 11 exhibition on July 20, 1969. Until 1972, the United States had carried out five human and several robotic expeditions to the Moon, and then the space race died down and with it also complicated, costly and risky expeditions. The last man left the Moon's surface on December 14, 1972.

The first soft landing on the Moon after the landing of the Soviet probe (Luna) on August 19, 1976, was carried out only by the Chinese probe (Čang"e 3) on December 14, 2013, and the last one (Čang"e 4) on January 3 2019 on to us invisible side of the Moon. In the meantime, in addition to the USA probes, the Japanese probes (Hiten) 1993 and (Selene) 2009), European (Smart 1) 2006 and Indian (Čandrajan 1)2008) also crashed controllably. At the moment, there are still a few landings at different stages of planning, including private ones. The next step in exploring the Moon with a human crew, with the establishment of more permanent settlements—is not yet in sight. Perhaps after the year 2025.

Picture 5 Geographical view of most landings on the Moon

Št.	Name of the expedition	Carrier rocket	Launch date	Landing date	Space program	Comments
1	Luna 9	R-7 Semjorka	31. January 1966	3. February 1966		First successful soft landing
2	Surveyor 1	Atlas-Centuar	30. May 1966	2. June 1966		
3	Luna 13	R-7 Semjorka	21. December 1966	24.December 1966		First probe with measuring instruments beside TV cameras
4	Surveyor 3	Arlas-Centaur	17. April 1967	20. April 1967		
5	Surveyor 5	Arlas-Centaur	8. September 1967	11. September 1967		
6	Surveyor 6	Arlas-Centaur	7. November 1967	10. November 1967		
7	Surveyor 7	Arlas-Centaur	7. Januay 1968	7. January 1968		
8	Apollo 11	Saturn V	16. July 1969	20. July 1969		First landing with manned crew
9	Apollo 12	Saturn V	14.November 1969	19. November 1969		Second landing with manned crew and return of Surveyor 3 probe parts to the Earth
10	Luna 16	Proton-K	12. September 1970	20. September 1970		
11	Luna 17	Proton-K	10. November 1970	17. November 1970		First use of lunar vehicle (rover Lunokhod 1)

12	Apollo 14	Saturn V	31. January 1971	5. February 1971		Third landing with manned crew
13	Apollo 15	Saturn V	26. July 1971	30. July 1971		Fourth landing with manned crew; first with Lunar Rover
14	Luna 20	Proton-K/D	14. February 1972	21. February 1972		
15	Apollo 16	Saturn V	16. April 1972	21. April 1972		Fifth landing with manned crew
16	Apollo 17	Saturn V	7. December 1972	11. December 1972		Sixth and the last landing with manned crew until now
17	Luna 21	Proton-K/D	8. January 1973	15. January 1973		
18	Luna 23	Proton-K/D	28. October 1974	6. November 1974		
19	Luna 24	Proton-K/D-1	9. August 1976	18. August 1976		
20	Luna 25	Soyuz	11. August 2023			Unsuccessful mission
21	Čang'e 3	Long March 3B Y-23	1. December 2013	14. December 2013		Landed on Sinus Iridum
22	Čang'e 4	Long March 3B	20. May 2018	3. January 2019		Landing on the dark side of the Moon
23	Čang'e 5	Long March 5	23. November 2020	16. December 2020		Chang'e 5 will be China's first sample return mission
24	Čang'e 6	Long March 5	May 2024			
25	Chandrayaan 1	PSLV – XL C11	22. October 2008	28. August 2009		Communication ist lost
26	Chandrayaan 2	LMV3 M1	22. July 2019			Orbiting without landing
27	Chandrayaan 3	LMV3 M4	14. July 2023			Successful landing by south pole

Table 2 All landings on the Moon

2.2. The beginning of the space race

In the 1950s, the superpowers Soviet Union and the USA were in the Cold War. They competed in nuclear tests and other military armaments. The Soviet Union took a further path, towards the space research. First, it launched a satellite, then a living creature, puppy Laika, and in 1961 a human, cosmonaut Yuri Gagarin.

Until then, the United States had had no way to show any significant achievement in the space research. In 1961, a rocket plane X15 was test-launched, which was attached to another aircraft. It was launched during the flight, and then the X15 reached Mach 6 speed.

After the Second World War, the Americans captured 100 German scientists and engineers along with the most famous Werner Von Braun. Werner Von Braun was the constructor of the famous V2 single-stage rockets and V3 two-stage rocket. These were the first ballistic rockets of the Nazi regime. The most striking rocket was V2, then the V3, which could reach the USA; however, this was not realized.

In the last days of the war, scientists escaped the penetrating Russians and surrendered to the American army, which also included 300 wagons of rockets and rocket parts for V2. Then they sent these scientists to the United States. This made Stalin so angry that he said: We defeated the Nazi army, occupied Berlin and Peenemünde, and the Americans were given rocket engineers." The Russians captured some outstanding experts and hundreds of lower technicians.

Together with scientists and rocket parts, 15 tons of documentation and plans were seized. Scientists later collaborated with the American space program because the USA did not have much experience in this field.

In the 1950s, the USA was a little lost in dreams because of comfortable living. The American consumer was not interested in anything but a job and good income, and that he could afford as much good as possible. Large production of household appliances, infinitely large cars, was taking place. Long and enjoyable holidays, however, were a dream of every American. Von Braun later said that in the years 1945 to 1951, Americans did not have any notable program for the production of ballistic projectiles. This six-year period when the Russians laid the foundations of their large rockets program was lost; for now.

Towards the end of the 50s, Americans had quite a few political defeats. From the failure at the Bay of Pigs, to the acquisition of the political influence of the Soviet Union in Asia, and the final touch were the successes of the Soviet Union in space. The Army instructed German experts to begin preparing plans for the production of rockets and training programs. They became American citizens and used their knowledge in projects for army projectiles in the Redstone arsenal. They built a Redstone rocket, which had a range of over 300 km. During that time, the Soviets made two moves, the first being the atomic bomb test and the beginning of the war in Korea. Both events prompted the Truman government that it began to issue money for weapons and modest funds for rocket projectiles of infantry and aviation. Land forces and aviation competed who will be stronger in rocket technology. Eisenhower's government, which tried to ease the battle between the two generations of the army, focused on the development of rockets. When the idea of military satellites appeared, the government decided that this scientific program should not hinder the decisive American race in perfecting rockets. The turning point occurred on July 29, 1955, when President Eisenhower announced that the Navy Research Laboratory would launch some satellites large as lemon, in the international geophysical year. Satellites were intended for scientific research only. The rockets that they were supposed to launch these satellites were known as the "Vanguard" project. With that decision, Eisenhower separated the satellite program from military rockets. The satellite program could actually push the satellites on the journey around the Earth, but with this, the president rejected Von Braun's plan to launch satellites and consciously pulled the line between US space-seeking efforts and military rocket programs that enjoyed the highest priority. One day after the Eisenhower's statement, the Russians reported that they also had plans to launch terrestrial satellites in the international geophysical year and said that their satellites would be larger than the American small balls "lemons". Von Braun and colleagues had been dreaming of travelling to the Moon for years and thought that this split was pointless and harmful. Scientists extended the Readstone rocket and planted ten small rockets filled with gunpowder. They thought that this extended version of the Readstone rocket would push a small "cluster" of rockets into nearby space, which would then fall off. Small rockets that had caught fire would have started up a virtual bullet far across the Atlantic Ocean. The rockets were inserted into sort of rotating bucket. This Redstone with rotating bucket was called Jupiter-C, and finally, on September 20, 1956, they launched a Jupiter-C rocket from Cape Canaveral, which flew 5000 km across the Atlantic and reached a height of 1000 km. If they had put a gun powder in the tip instead of sand, you could have pushed it even into Earth's orbit. They asked for permission, but they did not receive it from the government, so they missed the first orbiter of an object around the Earth. The rocket was out away though it could have been a response to the Russian provocation. On October 4, 1957, military rocket experts were convinced that Sputnik would achieve what they could not have by persuading the government.

Thus, on October 4, 1957, the Soviet Union launched the first satellite that orbited the Earth, and in the same year the Sputnik rocket with the first living creature in space – puppy Laika. This was a real shock for the USA. The same year the USA attempted to launch a satellite into space into the Earth's orbit but an accident happened. The rocket exploded and the Americans were defeated. The Soviet Union was progressing more and more, and the United States could not be satisfied with partial solutions that they tried to do with some unsuccessful launches of mini-objects in the size of a small ball.

On January 1, 1958, the United States first successfully launched the satellite into space, and discovered radioactive Van Allen belts. After several further failures, more so called Explorers were launched, which satellites have successfully orbited into the Earth's orbit. At that time, the Soviets had already launched Sputnik III, which was 56x heavier than all US satellites together. The thrust power of Soviet rockets was much greater, and they were already far ahead of the Americans. The US Navy, land forces, then began to actually compete in a space race. Big promises were anticipated; they wanted to send a man into space as soon as possible. President Eisenhower had to give in the pressure and on October 1, 1958 the NASA Space Agency was finally established. It was an independent agency that could compete in the space race with the then Soviet Union. Initially, this agency was only advisory—scientific, and after the Soviet successes, the Congress gave the entire space program under this agency. Overnight, some 8,000 staff members received a new master and a new program. In the end, experts from the Vanguard Naval Laboratory and the Redstone Army, who launched the first two US satellites, came under the same roof. The Sputniks, however, were nicely orbiting around the Earth. In December 1958, NASA launched a rocket Atlas into the Earth's orbit, which was the first intercontinental rocket of great power. By then, Americans had devoted a lot of funds to the development of ballistic projectiles but after the Sputnik, everything had to change because the USA was still in considerable lag. The tension was rising as they wondered what else Soviets were planning to do.

In very secret operations in 1959, the Americans launched on the ballistic flight the first living creatures—monkeys with the Readstone rocket. This was the forerunner of the later Mercury program with which they wanted to send the first man into space. First, two monkeys Rhesus Able and Baker were launched, and then on January 31, 1961, the monkey Ham, who was the predecessor of the first man's flight into space, using the same rocket and the same procedure as later for the first man's flight, Alan Shepard. Due to some mistakes in this flight, another chimpanzee, Ensis, who had flown twice around the Earth, was launched and the third round had to be cancelled as the carrier rocket collapsed. All monkeys survived. Today, however, it would be difficult to carry out such experiments because societies for the prevention of cruelty to animals have a major impact on public opinion. The Americans were not quite satisfied with the previous attempts, but disappointed by technical failures. On the seventeenth of July 1960, the Atlas rocket was launched by the Mercury ship; exactly the one that later brought the human to the orbit around the Earth, Shepard and Glenn. The rocket had been successfully launched, but later lost radio contact. When it disappeared from the eyes, it exploded and a big orange flame was seen. The irony is that NASA announced the Apollo project the same day, which was a project without money, ideas and contracts concluded. Prior to Shepard's flight, they had to test a rescue system with a combination of the Redstone rocket and the Mercury spaceship, but again the rocket failed to launch. Only the rescue rocket was successful in taking off. After this confusion and technical problems as well as attempts, they finally decided. Shepard goes into the spaceship and into the space, thus creating the Mercury program.

During the entire race that had occurred by that date, the straw that broke the camel's back was when the Soviets sent to the space a man, Yuri Gagarin, to the ballistic flight in April 1961, and at the end of 1961 they even sent him into the first orbit around the Earth. This was a hard blow against the Americans in a prestigious technology race. Newly-elected US President Kennedy called for a meeting of all competent politicians and technical experts. The president stated before all the people: "Can we manage the space in the same way as the Soviets, and if not, why not? What is needed for us to do what the Soviets can?" Since then, the budget of the civil agency NASA has only increased. The funds were available for all assigned tasks.

2.3. The Mercury program

The Mercury program received 1.8 billion dollars. On May 25, 1962, President Kennedy stated that the Moon was the goal; the goal was to bring the human to the Moon by the end of the decade. This is not so easy and as Kennedy stated it is very difficult, and that's why we will do it, he said. At NASA they started to work gradually and systematically. It's not only about putting a man on the Atlantis rocket and launching it into space.

The Mercury program was born. A Redstone rocket was used as a ballistic flight rocket, and the Atlas rocket for the orbital flight around the Earth. The first ballistic flight was done by Alan Shepard, who became the first American in space, followed by Grissom. At least three weeks after Grissom's success, when his cabin was sunk after landing in the ocean, the Soviets launched Major German Titus on the journey around the Earth. In the Vostok II spaceship, he orbited the Earth seventeen times. This Soviet success was proof of a great advantage. In fact, the Americans were still testing the Mercury program because they had quite a few technical problems with it. The flights had to be cancelled prematurely or they could not take off at all. The Atlas rocket had the only sufficient thrust for orbiter of spacecraft Mercury around the Earth. January 1962 was determined that John Glen, as the first American, would circle the Earth. The Americans wanted to find out all about this hero. John Glen was worried about the public's great interest. Before the flight, he was in a White House meeting with President Kennedy, who told him that it was not possible to separate the man from his mission and that the Americans had the right to know everything. On the day of the flight, the weather conditions were not the best, so the flight had to be postponed and finally carried out on February 20, 1962. That day almost everything stopped in America. They were all in anticipation of the biggest event. The rocket, a 125-ton monster, was loose from the ground, the lift was successful, and after a while the main carrier rocket went off. At that time, the first American was already in the orbit around the Earth. The inhabitants of Perth were waiting for a flying star when Mercury was flying over them. Glenn saw almost all the continents, and the lights of major cities. The spaceship was traveling at 28,000 km/h, experiencing many sunrises and sunsets in just one day. He circled the Earth 3x and safely landed with the help of braking rockets. The only thing that was worrying was that in the Mercury control center the light was burning, as if something was wrong with a protective coating that was protecting him while descending through the Earth's atmosphere at 4000 degrees Celsius. Despite this control light, he landed safely, and the sizzled from the heat in the Atlantic Ocean. It seems that America was relieved; the new hero was compared with Columbus and Charles Lindbergh. Congratulations came from all over the world, including from the Soviets. However, the Soviets stressed that it all happened 9 months after Gagarin. The Moscow radio broadcasted a comment condemning the US attitude towards Cuba, while Glenn was returning to Earth. The day after the flight, the Soviet Prime Minister Khrushchev and the cosmonauts praised Glenn's flight and even suggested that the country should join efforts and even propose a joint meeting of the two superpowers.

By then, the Americans had already launched 69 satellites and the Soviet Union 13 and now this—the Americans sent a man around the Earth. The technology giant woke up and it was decided. The President Kennedy honored Glenn for special merits. In this excitement, the president wanted Americans to send a man to the Moon before the end of the decade.

In 1960 NASA had 10,000 employees and 36,000 external cooperators, while in 1962 there were already 22,000 employees and 115,000 cooperators. In all this, money was also important. The space agency demanded 3.8 billion dollars, and so in 1963, together with all the expenditure, the value of the budget was about 5 billion. Four years after, they had been almost without resources. German rocket expert Werner von Braun was given a green light and all the available means that in 1962 he could start planning a rocket to travel to the Moon. The issues that awaited NASA have become more complicated and more expensive. If they wanted to send a man to the Moon, just one flight was not enough. After Glenn, three Mercury spaceships were sent around the world, Mercury 7, 8, 9.

In mid-August 1962, it was reported that the United States was in the lead, but on August 11, the Soviets launched Major Nikolaev with the spaceship Vostok III, and the next day the Vostok IV with Lieutenant Colonel Popovic. The spaceships had a purpose to dock with each other in space, they were 450 km apart, but that did not occur. Nevertheless, the cosmonauts talked between themselves in the space, and their spaceships weighed much more than the Mercury spaceship. The Soviets were still of great advantage, and both were involved in the Cold War; the rocket crisis in Cuba, which still prompted the space race.

2.4. Two years of waiting

With Cooper›s flight, the Mercury project ended and now the burden of progress has just been put on mechanics and engineers, flight experts and project leaders. These were people who were dealing with difficult and troublesome matters. Almost two years had passed that the next man flew into space. In the meantime, a tragedy occurred, and US President Kennedy died of the assassin's gunshot wounds. Nevertheless, the space program continued at same speed. There was a lot of construction in Cape Canaveral, on one side of a high building for rockets, hangars, warehouses, control centers, and on the other side there were hotels, residential areas, shops and a hospital. Overnight, Cape Canareval became an American rocket launcher. There were dozens of companies that were building there; the land value was a few times higher. Everything should have been happening in secret, but it was still such a big project that they could not hide it. In the summer of 1963, they dug out the first cave for launcher for Saturn 1 and 5 rockets. The space agency had to select the location of the flight monitoring center. They chose Huston because it was one of the decisive moments that they received a gift—1000 *morgens* of a land—from Rice University, and that Vice President Johnson and the President of the Space Council were both from Texas. The city of Huston began to grow, and thus expanded to almost reach the space center. In 1966 there were 5,000 employees in the center. By August 1963, more than 100 airlines had been established there, which were in constant contact with the space center. When the space complex was completed, it cost 312 million dollars in total and was ready to control spaceship flights. The Huston Center cost about a third as much as devices in the Cape Canaveral launch center, later Cape Kennedy. In 1968, NASA spent already 4.4 billion dollars on this infrastructure. Permanent inhabitants of the centers were future spacemen, young and old. They were much respected among the locals. They did not lack money; they had good cars and all the comfort. Shepard was also one of the elderly spacemen who became a millionaire at the age of 45, and John Glenn was even older. He thought that he could be among the ones selected for the project Apollo. People in charge somehow made him know that they would choose younger, more daring people, and that they should not risk their lives once again because they were already national heroes. Glenn was greatly affected by the murder of President Kennedy, as they were personal friends.

The next task for the USA was to walk a man and to dock two spaceships in the space. The Americans had to continue rapidly, the delays were not allowed to occur. The ultimate goal was the Moon, so they had to quickly focus on new programs, not just accumulating hours in space. In mid-1963, the Mercury project was closed, and by that the Gemini program was created.

2.5. Gemini Program

By mid-1965, the United States spent already 8.8 billion dollars on space flights with crew, and all the costs after Sputnik amounted to 14.4 billion. The space agency had already 33,200 employees and almost 377,000 people worked for it in other companies. The two-seater rockets Gemini were next in line, which could have been navigated, and by that, the orbit would have been changed randomly. In the meantime, other experts were already preparing the project Apollo for a three-person crew (described further below). Not all Americans were excited about the broadcast of these takeoffs, as they sometimes had to stop some sports broadcast. It seemed that people stopped being interested in space flights, just when they started winning in this. The Soviets also were not still at that time. On June 14, 1963, Vostok V flew into space and Vostok

VI two days later, led by the wife Valentina Tereshkova. The ships were flying barely 5 km apart, and the Americans wondered why they sent a woman who was not even a pilot. Therefore, it was assumed that the spaceships Vostok could not be steered from one orbit into another. Gemini, however, had this ability.

In the spring of 1964, the Americans successfully tested the model of the spaceship Gemini and Apollo, both without a human crew. In the autumn of the same year, the Soviets launched the spaceship Voshod I with a three-member crew. After this launch and after the sixteenth circulation around the Earth, the Soviets made a short break, as the crew reported the continuing nausea of cosmonauts in space. After six months, the ship Voshod II was launched, with the cosmonauts walking around the space, bringing the Soviets back into the lead. By the time the Americans launched the first Gemini spaceship, they had already launched 250 satellites and Soviet Union 112. Meanwhile, the Americans were consistently improving statistics in the ratio of successful against unsuccessful launches and had better statistics than Soviets. The Soviets had a three times stronger rocket than the American rocket Titan, which would bring the spaceship Gemini to space.

Not even a week after Aleksey Leonov's successful Soviet space walk, the Americans launched the spaceship Gemini 3 into space, which, after a one-year training, the spacemen Grissom and Young successfully navigated from a higher to a lower orbit and turned the ship to the left and to the right. The next turning point happened with the spaceship Gemini 4. On June 3, 1965, spaceman Edward Higgins walked around space in the third round. He walked with a space suit worth 28,000 dollars and at constant falling at a speed of over 30,000 km per hour without feeling any speed. He was looking at the black void, the twinkling stars, the moon, and the glittering Sun, and under him there was an aerial blanket that stretched over the whole Earth. For free movement around the space, he helped himself with an oxygen reaction pistol with which he also dragged the spaceship, thereby applying the law of action and reaction. US space developments also influenced Soviet flights, but White's and McDivitt's flight showed that sometimes also the United States imitate something. US space experts planned a walk around space only with Gemini 6, but after Leonov's walk they changed plans and included a 10 to 12-minute walk already with the Gemini 4. The US space program was also accompanied by incidents as they protested outside the space center against the segregation in Houston schools. There was also no shortage of protesters against the Vietnam War, and as a result they were later even reducing military space budget.

*Picture 6 Aproching Gemini
to Agena Picture*

*Picture 7 After merging
Gemini and Agena*

In August 1965, flights acquired a new dividing line. The smallest spacemen, Cooper and Conrad, had been circling around the Earth for eight days with the Gemini 5 spaceship. When the Gemini 5 landed in the ocean, the USA broke a new record, the longest flight with a human crew of 190 hours and 56 minutes, while the previous Soviet record was 119 minutes. Nevertheless, there was no particular enthusiasm for the record in America. Television stations were even accompanied by protests because they stopped sports broadcast during the takeoff. There was another story regarding Gemini 5 flight. Spacemen Conrad and Copper met a Soviet spacemen Leonov and Belyayev during Europe tour in Athens. They embraced themselves and together toasted to successes on both sides.

The next turning point happened with the Gemini 6 spaceship. This supposed to dock with a rocket that would have been sent to space before Gemini 6. This experiment was thought to have already been done with the Gemini 5 ship, but was abandoned due to errors. This attempt also failed because the "chased" rocket had exploded during the takeoff. In any case, they had to try to pursue another object on a circular path around the Earth because they needed a target for radar and a computer. They changed the flight program and launched the spaceship Gemini 7 with the spacemen Borman and Lowell. They used the Gemini 7 as bait for Gemini 6. The problem was that the same launcher had to be used for the launch. However, this launcher should, in normal conditions, have been prepared in a month for the launch of next rocket. Nevertheless, the ramp was prepared in eight days. First days in the space, Bormann and Lovell photographed and corrected the path with the short ignition of rockets. In the meantime, the crew of the Gemini 6 was training maneuvering on Earth, which was needed for subsequent chase with the Gemini 7 after the takeoff of the Gemini 6 spaceship. The takeoff of the Gemini spaceship had to be cancelled due to an error. The takeoff was successfully launched in a few days when the Gemini 7 was already circulating around the Earth. In less than four hours, the Gemini 6 set up a radar contact with the Gemini 7, which later switched to the Gemini 7 orbit. Soon they were 40 meters apart, and later they could see each other. This was the first planned successful approaching of the two US ships in space. At that time, the US had already launched twice as many satellites as the Soviets, but each year they lagged behind in the total weight of spaceships.

On Earth, in 1966, it seemed realizable that the USA would land on the Moon before the time it had set itself up. Gemini—an experimental spaceship for two crew members carried out the assigned task. In twenty-five months, twenty people were put into space and gave a lot more than seemed initially. This was a key test for the project of the flight to the Moon, during which it was necessary to learn on faults and successes how to fly the in the space. However, the idea was still far from reality, and in spite of all the successes so far, the Moon was seen as a picture in the sky and on a computer paper.

In Cape Kennedy, a takeoff named launch pad 39A for a flight to the Moon was built. In Huntsville, rockets were tested, in factories and laboratories at thousands of conferences they were bending over the maps of celestial paths and the plans of spaceships. On the twenty-eighth of February, the backup pilots Stafford and Cernan after the Gemini 9 became the first to be tragically killed as the main pilots. The accident did not slow down the realization of the program. Gemini 8 with pilot Neil Armstrong and Scott was preparing for the flight.

The task of Gemini 8 was to dock in the orbit around the Earth with the Agena rocket, which had previously been shot as bait for chase in space. After the chase drive, the Gemini 8 finally docked with Agena, which was full of fuel. This was the first time that two US spacecrafts docked with each other in space and flew together. It did not go smoothly, because the two docked spacecrafts started to rotate around the transverse axis. Rotation was faster and faster. Armstrong turned on the engines intended for returning to Earth so that he could move away from Agena, and a little later the pilots landed in the Pacific Ocean, as they used valuable fuel intended for return. The Agena rocket was lost, and for the next docking, they searched for another ATDA rocket, which was even more adapted for docking in the space. This new rocket was suitable for docking, separation and redocking. This was a turning point or essential element for the success of the Apollo program. The command module would have to dock with the lunar module "Spider", then to separate due to the descent to the Moon, and at the return redock again. Until the spacemen had completed this exercise, the Apollo program could not have been accomplished. Two days after the launch of ADTA rocket, spacemen followed it with the Gemini 9. When the spacemen approached the bait, they noticed an open lid on it, which was protecting it during the launch. The docking was wasted, and it was all glistening in the light caused by the snow-white clouds over the ocean. The spacemen became tired, and manoeuvring after the radar needed for approaching, cost a lot of fuel. They moved everything on the second day in which they practiced a walk in space. This walk also proved to be a great struggle. All together, it took a lot more time and effort than they imagined. Difficult breathing and freezing of the breath on the pane. Stafford, however, was walking in space for two hours, which is the most so far. When landing on the Earth, the Gemini 9 helped itself with the ship's computer better than any spaceship so far.

The Gemini 10 was waiting in line, and in July, pilots Young and Collins flew towards the space, where two ships had been waiting for them. The first was lost Agena from the Gemini 8 and the new Agena, which was launched precisely for the Gemini 10. In the fourth round, they caught it, docked with it, switched off their engines, and turned on the Agena's. This power took them to a height of 766 km, which was a new world record. Then they disconnected from this Agena and approached the old Agena which had been orbiting the Earth for eight months. Collins had to take some instrument down from it and get back to his spaceship and disconnect from a 17-meter-long tube through which the oxygen for breathing was obtained from the mother ship. This flight was a fine success, full of rich experiences of approaching and docking of two ships in space.

By the end of this program, two flights were foreseen. The USA was already far ahead of the Soviet Union. It had carried out fourteen flights with the crew, the Soviets eight. USA had launched 22 spacemen, and the Soviets eleven. American spacemen were on the way around the Earth 1661 hours and 52 minutes, which is three times more than the Soviets. American spaceman accumulated for two hours and 56 minutes of space walks, while the SU had only Leonov with ten minutes. Finally, the USA had seven approaches and two dockings behind itself, and the Soviets had none.

The Moon was shining brightly at that time, and the spaceship Surveyor landed lightly on her, sending hundreds of photos to the Earth. This was the beginning of the exploration of the Moon because it was the best place for landing of the first spaceship with a human crew.

Not everybody on Earth was aware of what was going on and on what journey mankind descended. There was poverty throughout the world, and some people doubted that the right things were happening with this program. Give or take arguments, in 1966, the space program cost more than any year before, nearly 6 billion dollars. The space program has also reached a peak in employment. 34 000 people were working for it directly and 360 000 people indirectly. The peak of the preparations for the flight to the Moon was achieved and the flight to the Moon could not have been replaced by any equally important goal. In the meantime, the Vietnam War was raging, which cost more and more year after year, and with that some savings were also made on the space race.

The Mercury project proved that the space flight was possible, and the Gemini project proved that a man could play a much larger role than he had imagined when flying around 30 000 km per hour and bound by the law of circular orbits but sufficiently free that he could move in the sky at his own will. Apollo and the following programs took advantage of both.

The launch of Gemini 11 with the crew Conrad and Gordon was fixed for September. Spacemen wanted to set a record and catch up and dock with the Agena rocket already in the first round. With the help of the radar they found it and docked with it. It was not just a speed exercise but also a manoeuvring exercise that the spacemen would have to do when the <spider> docked with the mother ship when it returned from the surface of the Moon. The next day, Gordon got out of the cabin and headed for the first space walk, tied to a 10-meter-long breathing tether. During the attaching, he was quite out of breath and began to sweat, so he had to stop the walk prematurely. Later, docked with Agena, they used its engines and flew with its power 1368 km high. This was the world record. They saw almost the entire Earth at this height. For some time, they were still docked with Agena only by a tether that they literally pulled with them. With the explosive device, they disconnected the tether in the rocket's nose and separated from it. The next day, they approached Agena again, but they did not test the second part of the task – to do work in space. The spacesuit still looked very cumbersome and it limited free movement so that a spaceman became very tired during the movement. It always lifts you up and carries you, so the tasks must be simple and easy, the spacemen established. Gemini 12 was the perfect descendant of Mercurys, and the last ancestor of Apollos. On November 11 1966 it was steered by Lowell and Aldrin. After a few hours after takeoff, they caught the Agena rocket and docked with it, but due to engine problems, they had to separate from it. They performed the longest trip outside the ship until then, photographed the Moon, the Milky Way, and the Earth. The flight with the Gemini 12 was a complete success due to the length of the walks taken and the work done outside the spaceship. This proved that spacemen can move and work in an inhospitable environment. They landed in the Atlantic Ocean, very precisely, near to the rescue ship. It was the last stage in the Gemini program that produced and tested methods as well as tools that helped the Americans in the path of conquering the Moon. They did everything they needed to make the next step, the Apollo program, which today is still considered to be the most complicated of all times. Following these successes, President Johnson addressed the Americans; "Now America will remain in space".

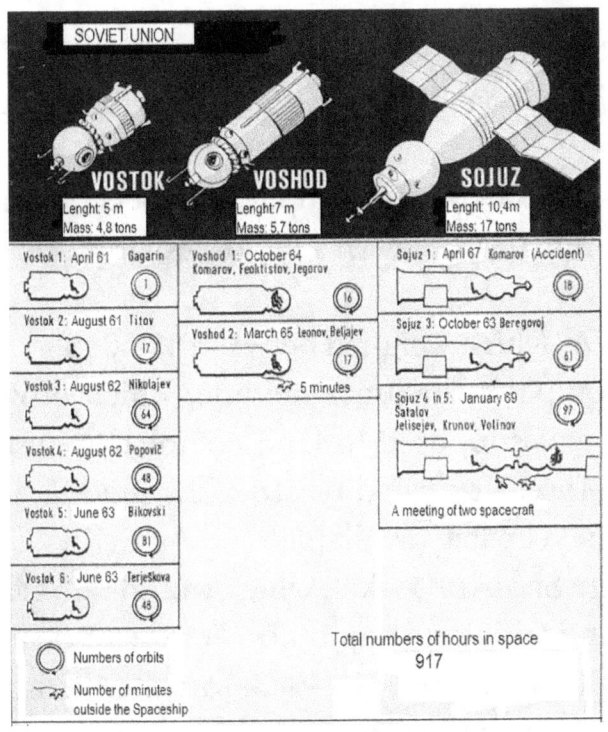

Picture 8 Soviet Space travels

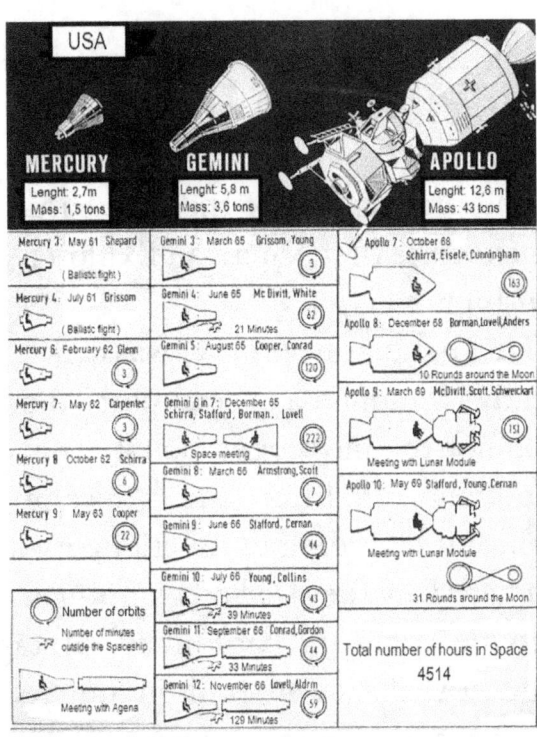

Picture 9 USA Space travels

2.6. The Apollo program

2.6.1. Foreword

The first flight of the Apollo spaceship was scheduled for January 1967, two months after the Gemini 12. The crew was selected in March; these were Captain Grissom, the veteran of the first flights from the Mercury and Gemini program, and the first American spacewalker, Ed White and novice Roger B. Chaffee.

The Apollo program has always been the most complex project in the history of flights into space. For the launch vehicle of the flight, the most powerful rocket of all time was used (Saturn 5). The program is a multi-complex system with many subsystems and flight phases. It is a rocket consisting of several thrusting modules, each of which has its own task to perform a flight on an exactly specified orbit. It is impossible to just build a rocket that would take off from Earth, land in one piece on the Moon, and then fly back to Earth. For such a task, the rocket would require a lot of fuel and would weight too much.

2.6.2. Brief description of predevelopment of the Apollo program

The Apollo program was entrusted to the project leader Werner von Braun. The first plans for rockets for the flights to the Moon began in 1961, even before the Gemini project.

The predecessor of the first Saturn test rocket was the Jupiter C rocket, which launched the first American Explorer 1 satellite in 1958, with a ten times larger thrust than Jupiter-C.

The first planned rocket was called NOVA, which weighed between 45-50 tons and already contained a special part—the prototype of the lunar module, which proved to be not the most operational. That part was supposed to land completely on the Moon, and lifted from the surface of the Moon. It turned out that the module was too heavy according to functionality and fuel capacity available.

That's why they started planning a 3-stage Saturn 1 rocket, which would be able to fly to the Moon and back in several stages and with larger modules. For the tests, four Saturn 1 rockets were built in four years (Chapter 2.6.3.). In 1963, together with the Mercury and Gemini projects, 375,000 people were responsible for this development.

For the development of computer equipment, NASA signed a contract with IBM. Nobody knew whether the development of computer equipment would be done fast enough and if the software would reach the required level to enable the operation of such a complex system as the Apollo program.

Saturn 1 rockets were successfully tested, but the final model of the rocket was renamed Saturn 5, as the rocket was more powerful and upgraded with several additional modules. Among other things, a two-part lunar module was created, which enabled a safe landing and takeoff from the lunar surface. The both parts were supposed to land on the Moon, and the takeoff would take place only with one part, while the other would remain on the surface of the Moon.

Programs should follow according to the concept and development stage: Mercury, Gemini and Apollo. Due to the time limit and the space race, the development was carried out in parallel with the Mercury and Gemini programs, so that twice as many people as they needed were employed, in a different sequence. Also, rockets for each program were developed by different teams, from Atlas, Titan and Saturn. With this, they gained a lot of development time, except that the team that had experience from one program could not transfer the experience to another program. In that exceptional time everything was at hand, money, will and people. One of the most demanding parts in the development of Saturn 5 was the F1 rocket engine, which prototype was created in 1950s and in the last version it had 620 tons of thrust. They used even five engines for Saturn 5 for the first stage. Altogether, the rocket was given 3300 tons (7.6 million pounds or 35.1 M Newton) thrust at the takeoff. These engines thrusted the rocket of nearly 3,000 tons up to 8000 km/h with acceleration 4G.

Picture 10 Assembly 5x
F1 engine (Saturn V)

Picture 11 Size of Engine F1

At the beginning of the development, the engines had some flaws. The burning of three tonnes of kerosene and liquid oxygen per second was unstable, resulting in vibration along the entire rocket. This reduced the ability to properly increase or decrease power in the takeoff and flight of the rocket. Thousands of engineers were working on this case, which was later solved successfully with additional injector nozzles. Rockedyne's F1 engine has become manageable and stable.

2.6.3. Predevelopment of the Apollo Saturn rocket program

The development of the Apollo program began very early, in 1961, together with the Mercury program. According to the concept and development stage, programs should follow as stated: Mercury, Gemini and Apollo.

Due to the limited time and the space race, it was carried out in parallel with the Mercury and Gemini programs. Also, rockets for each program were being developed by different teams, from Atlas, Titan and Saturn. With this they gained a lot of development time, but they needed several times more money and people. In these exceptional times everything was available.

The forerunner of the first Saturn test rocket was the Jupiter C rocket, which launched the first American Explorer 1 Explorer in 1958, only that it had a ten times greater thrust than Jupiter C.

SA-1

Nasa wanted to test each stage individually. First, the S-1, which was carrying also two dummy upper stages, and the SA-1 with which they wanted to test eight H-1 engines and a stage, as well as S-1 stage with a fuel tank and a control device. Rocket twisting, shaking, and pressure measurement were tested. The first flight was launched on October 27, 1961, and it was going without problems, reaching a height of 136.5 km and giving 485 measurement data. It was the first unmanned flight of Saturn I.

Picture 12 SA-1 Rocket

SA-2

It was the same as SA-1. It got new engine and was carrying two dummy upper stages. The flight was ballistic without a specific goal. That year they tested the first stage, its structure, control, launch devices, aerodynamics and tested the release of 109 thousand liters of water into space (Project Highwater 1). The rocket was launched on April 25, 1962. At a height of 109 km, water was released that formed a cloud of

4.6 miles and climbed into the ionosphere at a height of 160 km, thus measuring the influence of the oxygen, hydrogen, and water at this height.

Picture 13 SA-2 Rocket

The SA-3 was similar to the SA-1 and SA-2 missions, carrying the maximum amount of fuel. It was carrying the empty second stage of the rocket and the head from the Jupiter-C rocket and did not have an active guidance. It also released 109 thousand liters of water (Project Highwater 2). The poor telemetry of the flight caused doubt in the accuracy of the data obtained. Minor accelerations were tested on the rocket and thus longer fuel burning. It reached a height of 167 km.

Picture 14 SA-3 Rocket

SA-4

The mission was similar to previous missions, with the difference that they did not carry water. During the flight, they carried out a test by separating one of the engines to determine how the rocket behaved in such a case. This experiment was later used in the Apollo 6 and 13 programs. The launch took place on March 28, 1963. The mission was the last one to test only the first stage. Instead of an empty second stage, an aerodynamic imitation of second stage was used on this flight. During the ascent, one of the engines was shut off, and the fuel was rerouted to the other engines. The rocket reached a height of 129 km at a speed of 5906 km per hour. Also on this mission, retrorockets were used to separate the individual stages. They did not separate the stages, but only tested their operational efficiency and readiness. The mission was completely successful

Picture 15 SA-4 Rocket

SA-5

This mission was the fifth launch of Saturn I and the first of Block II, which represents a more advanced rocket configuration. It had a live second stage and devices for self-guidance of the rocket. For the first time, the Saturn I would fly with two active stages—the S-I first stage and the S-IV second stage. At the top, the rocket had a Jupiter-C nose filled with sand to simulate the weight of the lunar module. With this flight they also tested the control system, the active second stage, the first and second stage separation, the functionality of the launch site, the operation of the TV camera, experiment with liquid hydrogen on the second stage, the eight holders at the launch site and the aerodynamic fins for stabilization at the first stage, and the flight in orbit with 17 tons of cargo. The flight was launched on January 29, 1964. After separation of the first S-I, deviations occurred that affected the active second stage, which deviated by 21 km in the flight. It reached perigee of 262 km and the apogee was 785 km. The engine thrust was 836 KN, and TV cameras were used which recorded the behavior of the rocket at takeoff, and during the successful flight they obtained as many as 1183 data.

Picture 16 Test model of the command modul (boilerplate)

Picture 17 SA-5 during the launch

SA-6 (A 101)

It is a mission in which the first model of the command and service module (boilerplate) was used, (therefore called A 101), in order to establish the compatibility with Saturn I. The goal of this mission was to test the rescue capsule with the rescue engine. The structure of the SA-6 was therefore the first and second stage, the rescue system- launch escape system (LES)—and the imitation of the command module, which, by mass and shape, was similar to the Apollo spacecraft module. Takeoff was on May 28, 1964. Telemetric data showed that aerodynamic heating was 20% below the permitted limit, the first and second stages were successfully separated, and the rescue system was activated. The command and service module was orbited to a height of 182 x 227 km and remained there only 54 circulations around the Earth. The reason for this was the failure of the engine equipment with the turbo pump. For the following flights, engineering corrections were made.

Picture 18 SA-6 (A-102) Rocket

SA 7 is exactly the same mission as SA-6, which took off on September 18, 1964. Heating during the takeoff was normal. The difference between A 101 and A 102 was that A 102 had an instrument for measuring temperature fitted on the command and service module. Launch escape system was successfully tested. After activating the engine, the capsule was successfully separated from the rest of the spacecraft. For the first time, they successfully used a computer on the Saturn I that could be reprogrammed during the flight. After 59 circulations around the Earth, the module successfully landed in the Indian Ocean. The mission was completely successful.

Picture 19 SA-7 (A-102) Rocket

A 103

It was the eighth flight with unmanned crew. The rocket system was exactly the same as for A 102, but A 103 launched into the orbit the Pegasus satellite which collected data on meteorites in the Earth´s orbit and was located in the service module.

During the launch, they tested various assemblies and activated the rescue module of the spaceship. The ship reached an apogee at 743 km and a perigee at 495 km.

Picture 20 A-103 Rocket

A 104

This mission launched the second satellite Pegasus B. The goal of this mission was to collect data on the meteorite's activity in orbit and to show the accuracy of the launch rocket system. Saturn I and its cargo were similar to those of the A 103, except that another independent engine and instrument for measuring temperature during takeoff was added to the service module. There was another difference, two engines in the Apollo spacecraft were namely a prototype, and the other two were simulation. The total mass of devices launched in orbit was more than 15 tons. Pegasus B was retired from operation on August 29, 1968.

Picture 21 A-104 Rocket

2.6.4. Development of the safety system (Launch Escape System)

Takeoff of the rocket is one of the most dangerous parts of the flight. For this reason, they were developing a safety system that would allow the rapid evacuation of astronauts in case of danger. The command module, in which the astronauts were located, was connected to a small rescue rocket, which was installed at the top of the entire rocket.

Test take off

The first test was done on June 28, 1963 at the location of the new New Mexico by the imitation of the command module Little Joe II. It was a test flight of the first stage on which the inactive command module and rescue system were inactive. They reached a maximum height of 7 km.

Abort of the mission at the launch ramp - number 1

The second test was carried out also without a human crew in which the command module and rescue system (Launch escape system LES) were active. The test was carried out on November 7, 1963; 15 seconds after the main rocket took off, with the ignition of the rescue system engine. The command module successfully landed with parachutes, only a slight instability of the rescue rocket was detected.

A - 001

This mission was intended to abort the flight at a high dynamic pressure level. Seven engines were used on the system, one with 42-second thrust, and the others with a 1.5-second thrust after ignition. The takeoff was successful on May 13, 1964. The signal from Earth interrupted the flight of the launch vehicle 44 seconds after taking off at a height of 9.1 km. The system successfully separated itself from the launch vehicle and returned to Earth with a parachute. In this case, one of three parachutes failed.

Picture 22 BP-6 (Boilerplate) with launch escape system(LES)

A - 002

This mission was similar to A-001. Its goal was abort test of the takeoff at altitudes where there is high pressure; with the exception that the controlling instruments were installed on the launch vehicle, better parachutes were built-in, and also fins were positioned for orientation and stabilization on the rescue spacecraft. The command module had additional protection against the exhaust gases from the rescue rocket. The flight was on December 8, 1964. After the flight had been aborted after 11 seconds, the rescue rocket separated, then it was spinning, and after a while it stabilized. The system reached a height above 15 km; the parachutes were successfully activated and allowed a soft landing of only 7 meters per second.

A - 003

With the fourth test A - 003, they tried to test the abort at high altitudes again. Takeoff occurred on May 19, 1965. The spacecraft did not reach the estimated height of 37 km, since it moved away from the control 2.5 seconds after takeoff. The carrier rocket fell apart before the second stage switched on, and started to spin 260 degrees per second and the rescue rocket with 20 degrees per second. Later it turned out that the fins on the rescue rocket were not functional. Nevertheless, the command module successfully landed on Earth with parachutes. The heat shield passed the test and was used as such on Apollo missions.

Abort of the mission at the launch ramp - number 2

This was the second abort test of the mission on the takeoff pad where the command and service module from A-002 was used. After the takeoff on June 29, 1965, the rescue rocket began to rotate around its axis, which did not affect the positive result of the test. The rescue system and the heat shield on the command and service module were successfully separated from the module. This module also had glass built-in for the crew that successfully passed the influence of the exhaust gases.

A - 004

This test contained a prototype of the command and service module for the Apollo program for the first time. It was the fifth and last flight of the Little Joe II rocket, launched on January 20, 1966. After the height reached and 2.9 seconds, the abort of the flight was activated. The rocket was rotating at 160 degrees per second, and because of the fins on the rescue system, the command and service module was successfully stabilized. The maximum height reached was 23.8 km. The parachutes opened and the command and service module successfully landed 34.6 km far from the takeoff site.

Picture 23 Litlle Joe II, misija A-004

2.6.5. Saturn 5

After years of planning and testing, the final form of the Saturn 5 rocket that flew people to the Moon was created.

Picture 24 Development: Size between Atlas and Saturn 5 Rocket

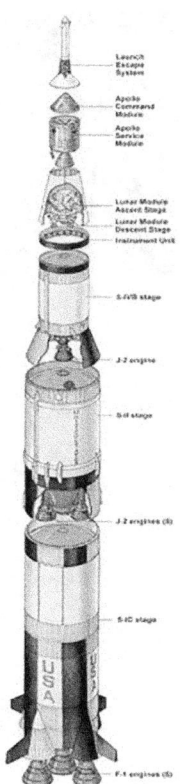

Picture 25 Saturn 5 Assem

Basic data for Saturn 5 Rocket

Function: single-use heavy carrier rocket
Manufacturers: Boeing (S-IC); North American (S-II); Douglas (S-IVB)
Country: USA
Height: 110.6 m
Diameter: 10.1 m
Mass: 2,800,000 kg (2,800 tons)
Stages: 3
Capacity: in LEO 118 000 kg; on the Moon 45 000 kg
Status: no longer in use
Launch: LC-39, Kennedy Space Center, Cape Canaveral
Total launches: 3
Successful launches: 12
Partially successful launches: 1 (Apollo 6)
First launch: November 9, 1967 (SA-501)
Last launch: May 14, 1973 (Skylab)

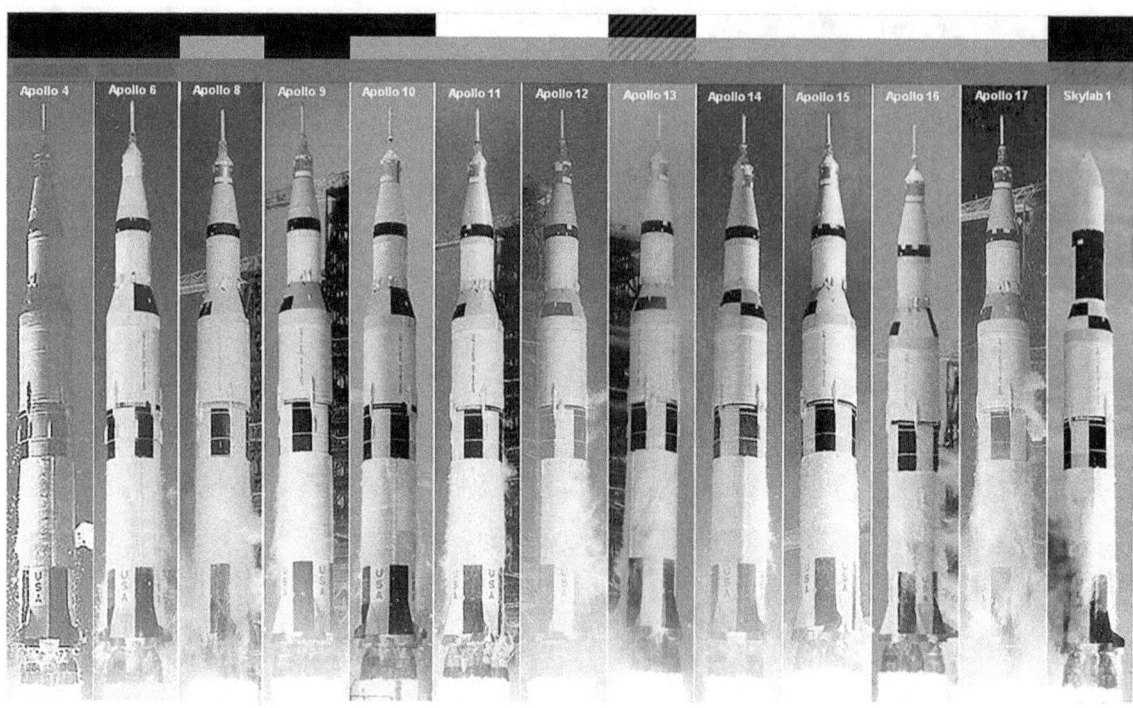

Picture 26 Saturn 5 rokets (Apollo 1 to Skylab1)

1. The first stage: ascent of the rocket to the first orbit (separation from the first stage)

This stage is the first and the most powerful. It is responsible for takeoff of the current maximum total mass of the entire rocket (3300 tons), up to a height of 68 km and acceleration 4g to a speed of 2, 75 km per second. When it expends fuel, it separates and falls into the ocean.

First stage - S-IC
- **Length**: 42.1 m
- **Diameter**: 10.1 m
- **Empty mass**: 131,000 kg
- **Gross (loaded) mass:** 2,300,000 kg
- **Engines**: 5X Rocketdyne F-1
- **Thrust**: 7,648,000 pounds (34,020,000 N)
- **Specific impulse**: 263 s (2.58 km/s)
- **Burn time**: 150 s
- **Fuel**: RP-1 / LOX (kerosene and liquid oxygen)

2. The second stage: ascent of the rocket to the second orbit (separation of the second stage)

The second stage lifts off the rest of the rocket to a height of 100 miles (176 km) and sends the spacecraft onto the orbit around the Earth. It accelerates the rocket to a near-orbital speed, which is 7 km per second.

Second stage - S-II
- **Length**: 24.8 m
- **Diameter**: 10.1 m
- **Empty mass**: 36,000 kg
- **Gross (loaded) mass**: 480,000 kg
- **Engines**: 5X Rocketdyne J-2
- **Thrust**: 1,000,000 pounds (4,400,000 N)
- **Specific impulse**: 421 s (4.13 km / s)
- **Burn time**: 360 s
- **Fuel**: LH2 / LOX (liquid hydrogen and liquid oxygen)

3. The third stage for the propulsion of the rockets from the Earth's orbit on its journey to the Moon

The third stage is the one that gives the acceleration to the spacecraft to separate from the Earth's gravity for the path to the Moon.

After the second stage was separated, the third one ignited twice. First, to accelerate the rocket to 7.91 km per second, this is at orbital speed. Then the rocket makes 2.5 orbits around Earth, so called parking orbit. During this time, astronauts check if the systems that are responsible for the start of the journey towards the Moon are working. Secondly, it ignites to accelerate the rocket to 11.2 km per second for the journey to the Moon. There is a vacuum in the space, where there is no friction and air resistance, and at this speed earth gravity is overcome.

Third stage - S-IVB

- **Length:** 18.8 m
- **Diameter:** 6.6 m
- **Empty mass:** 10,000 kg
- **Gross mass (loaded):** 120,800 kg
- **Engines:** 1X Rocketdyne J-2
- **Thrust:** 225,000 pounds (1,000,000 N)
- **Specific impulse:** 421 s (4.13 km/s)
- **Burn time:** 165 + 335 seconds (2 ignitions)
- **Fuel:** LH2 / LOX (liquid hydrogen and liquid oxygen)

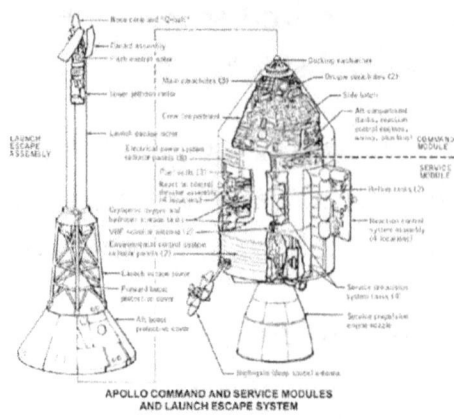

Picture 27 CSM and escape System

4. Separation and rotation of the CSM and the connection of the spider (LM), which was in the third stage

At the top of the third stage, a lunar module (LM) is stored in a special aluminum case, followed by a 3rd stage. CSM (Comand and Service module) with its own propulsion is separated from the casing where the lunar module is „garaged"; it rotates, connects the LM, and pulls it with itself. The Comand and Service module (CSM) and LM, that are connected rotate and jointly fly towards the Moon.

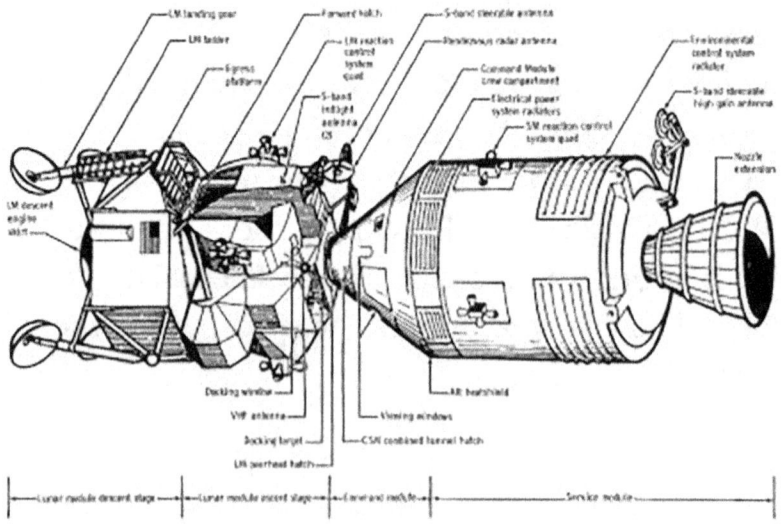

Picture 28 CSM + LM (Command-service and lunar module)

5. Orbiter of the CSM + LM module into the Moon's orbit

The CSM includes the braking engines that work for a few seconds to decelerate (CSM + LM) to the lunar orbital speed. Thus, CSM + LM orbits into the lunar orbit.

When the module is orbited into the lunar orbit, a process begins in which two astronauts move from command to lunar module

6. LM landing on the lunar surface

The LM and CSM are separated and fly together for some time. The lunar module ignites the braking engines, starts with a deceleration of speed and parabolic descends onto the lunar surface. It consists of two parts, landing and takeoff. Both parts have fuel for correction nozzles when landing and taking off. The landing module has its own engines and fuel for landing.

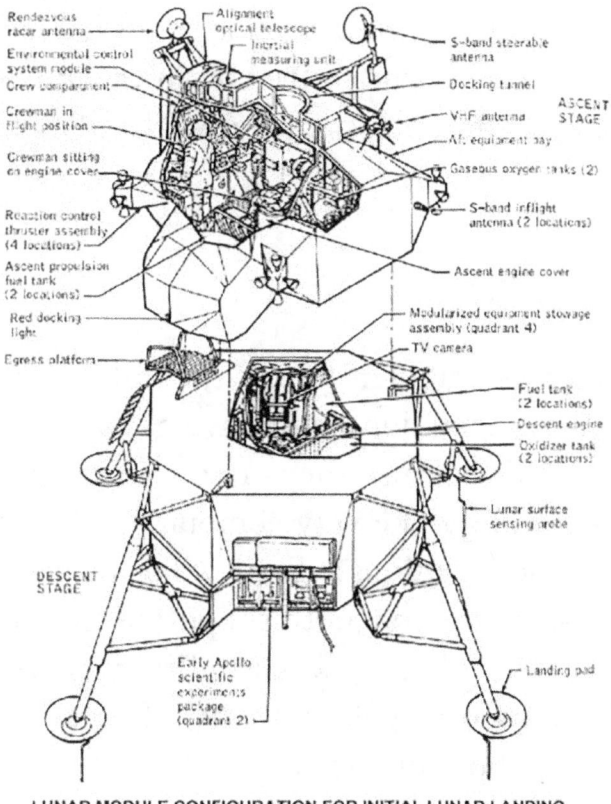

LUNAR MODULE CONFIGURATION FOR INITIAL LUNAR LANDING

Picture 29 Lunar module (LM) configuration

7. Takeoff of the lunar module from the lunar surface from the landing part of LM and docking with the CSM

After the ignition of the thrust nozzles, one part of the lunar module with the astronauts ascends from the lunar surface. Another part of the LM serves as a takeoff pad, which, however, remains on the lunar surface. During this time, CSM circulates around the Moon with a third astronaut. LM is later connected to CSM in the orbit around the Moon. The astronaut moved to the CSM from the lunar module. The LM is separated from the CSM and is propelled towards the Sun.

8. Returning

The three astronauts are with CSM that has the task to ignite the engine, and to deorbit from the lunar orbit and fly towards the Earth. Before entering the Earth's orbit the Service module separates from the Command module. The service module flies towards the Sun, and the Command module enters the Earth's atmosphere. After a large braking of the spacecraft, due to friction in the atmosphere, the Command module decelerates the speed to open parachutes that allow the three astronauts to safely land in the sea.

2.6.6. Apollo Computer System

This chapter describes a computer system that allowed people to travel to the Moon. The people of this generation, who have grown up with a computer, find it difficult to understand this because they are convinced that the computer age is typical of the present time only. The very computer technology, developed for the Apollo program, has provided solutions without which we cannot imagine a computer today. Its hardware and software were both simple and easy to understand.

Apollo Guidance Computer (AGC) was the first modern computer to perform tasks in real time. It was used to collect flight information and to automatically control all the Apollo navigation functions.

For example, the first Mercury spacecraft flights were controlled and guided by a ground radar and by the spacecraft radar controlled by the astronaut. Flights were made almost at random, with no automatic guidance. In accordance with the radar data, the astronaut controlled the rocket motors. In the Geminni program, an analog computer was already installed to control the docking of two spacecrafts.

The scientists knew that by using such navigation equipment they were not able to send a man to the Moon. Therefore, in 1961, NASA commissioned the Massachusetts Institute of Technology's Instrumentation Laboratory, later renamed as the Charles Stark Draper Laboratory, to create a digital computer which was produced by the Raytheon Company. The Massathusetts Institute of Technology (MIT) was assigned the task of putting together hardware and software for all the scheduled flight stages. Margaret Hamilton, the head programmer at MIT, was responsible for the software. The task of the digital Apollo Guidance Computer (AGC) was to provide guidance, navigation and control of the Apollo spacecrafts. Such tasks were tackled by outstanding experts. 2000 years of engineer hours were spent for development, that is, as if only one engineer were developing the AGC for 2000 years, however, there were up to 350 engineers who were working simultaneously. NASA was not familiar with all the required objectives either. And programmers did not know 100% of how they could complete the task.

This was the time when the integrated circuits (microchips) were used for the first time. In 1964, one microchip was worth $ 1,000. In 1963, MIT used 60% of all the American microchip production (100,000 pieces). The price had already fallen to $ 25 per piece until 1964.

	Block I	Block II
Computer type	IC Microchip Apollo 1- 6 (1964)	IC Microchip Apollo 7-17 (1966)
Memory	ROM 12K, RAM 1K	ROM 36K, RAM 2K
	16 bit (15 data+1 parity bit)	16 bit (15 data+1 parity bit
Tact Crystal	2.048 MHz	2.048 MHz
Gate	IC single 3-input	IC double 3-input
Logic	5000 ICs (3-input NOR gates, RTL	5000 ICs (3-input NOR gates, RTL
Processor	16 bit (14 bit + 1 stream processor + 1 signal processor)	16 bit (14 bit + 1 stream processor + 1 signal processor)
Operating system	EXEC	EXEC
Power consumption	55 W	55 W

Table 3 Properties of the Apollo Guidance Computer (AGC)

The AGC is the most interesting early computer because:

a) it flew the first man to the Moon, and
b) it is the first computer with the integrated circuit technology (microchip)

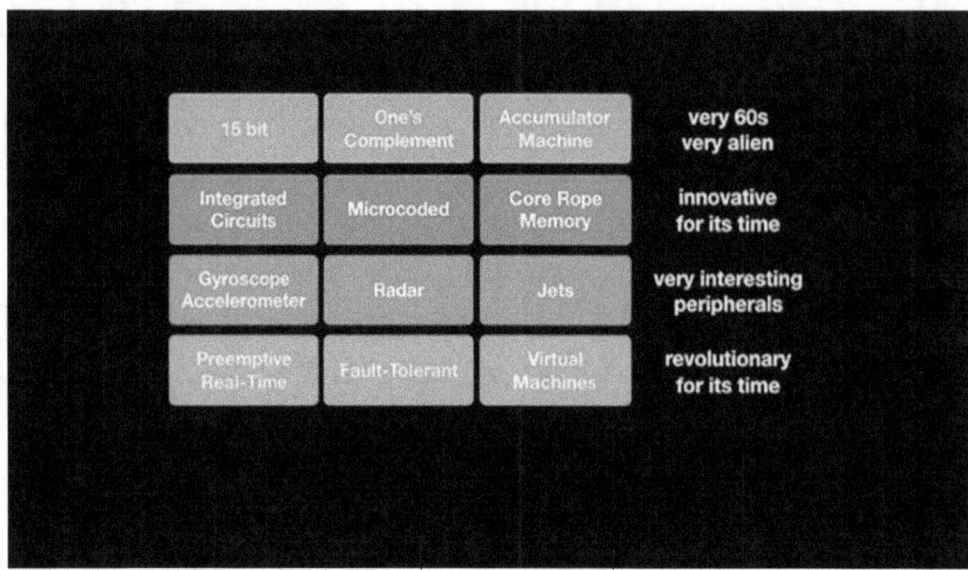

Picture 30 The AGC features

How did the Apollo flight computers get men to the Moon and back to the Earth?

The most common doubt that the conspiracy theorists have about the reality of the Moon landing refers to the computer capabilities, or to the question of how the computer of that time was able to navigate the Apollo ship from the Earth to the Moon and back. They are constantly giving us an example of today's smartphone which is able to capture much more data in a much smaller space, thereby indicating that it is also a hundred thousand times faster. This is a totally wrong way of thinking because it gives rise to wrong ideas about the capabilities of the computer that was made for this mission at that time.

There were 4 computers available to Apollo for carrying out its mission

1. The **Launch Vehicle Digital Computer (LVDC)** was designed for launching to the Earth orbit. This was the computer that was 40 years ahead of its time (see picture as evidence). The picture shows the size of the computer printed circuit board for the Gemmini program. A few years later, a printed circuit board was developed for the computer that controlled the Saturn 5 rocket liftoff. This printed circuit board has the capabilities of the printed circuit board that was used 40 years later in exactly the same available space size.

2. The Apollo Guidance Computer (AGC) with Display Keyboard (DSKY)

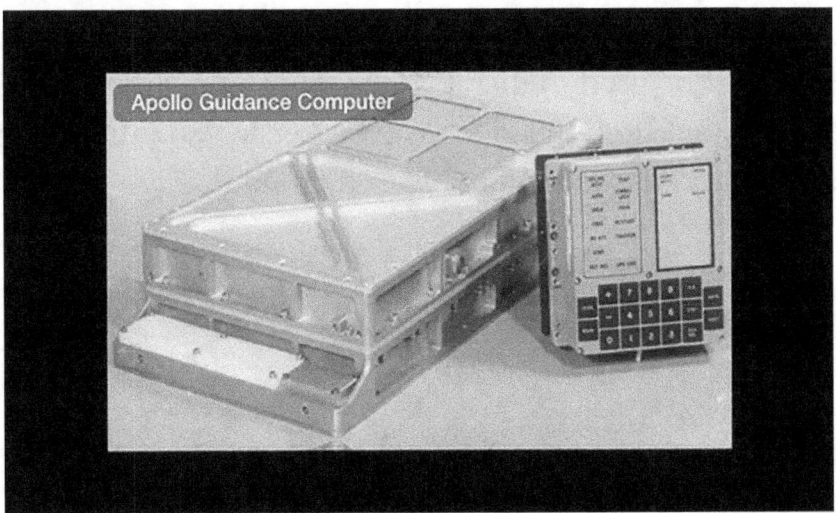

Picture 31 The Apollo Guidance Computer (AGC) with Display Keyboard (DSKY)

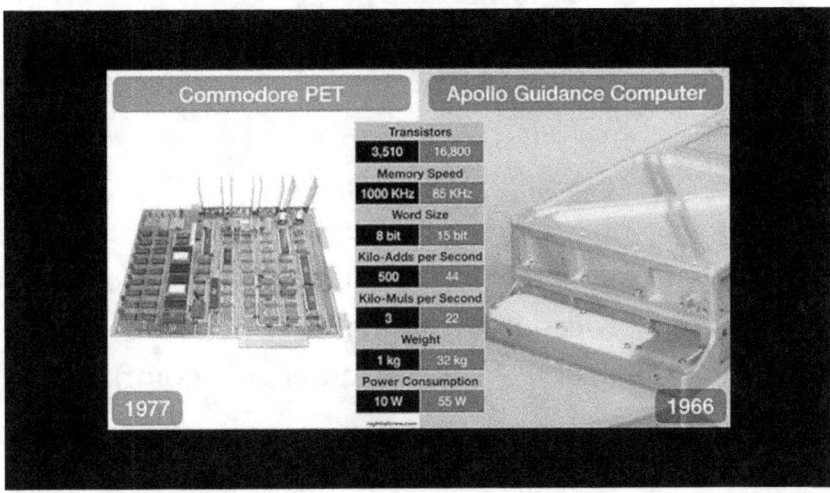

Picture 32 A comparison of the computer in 1977 with that in 1966

There were two AGC computers, one was in the command module for navigating the spaceship from the Earth orbit to the Moon orbit and back to Earth. The other one was in the Lunar Module. It was designed for enabling the lunar landing module to sit on the Moon's surface, and for launch in the Lunar Module back to the command module. The basics of commands were navigation parameters and the stored pieces of software that formed the whole of corresponding control commands for control devices.

AGC had no monitor mouse. It was designed to perform specific tasks. It was logical and easy to read, and without unnecessary interfaces. Already here we can see that the computer was optimally designed because it needed no capability for unnecessary space occupancy in the computer operations.

How did the astronauts communicate with DSKY?

DSKY human interface was built around the concept of Verbs and Nouns. For example, Verb 06 could be used to display the values of nearly 100 different memory locations selected and based on the given Noun. Every Verb and Noun had a specific task. The basic command list was available to astronauts.

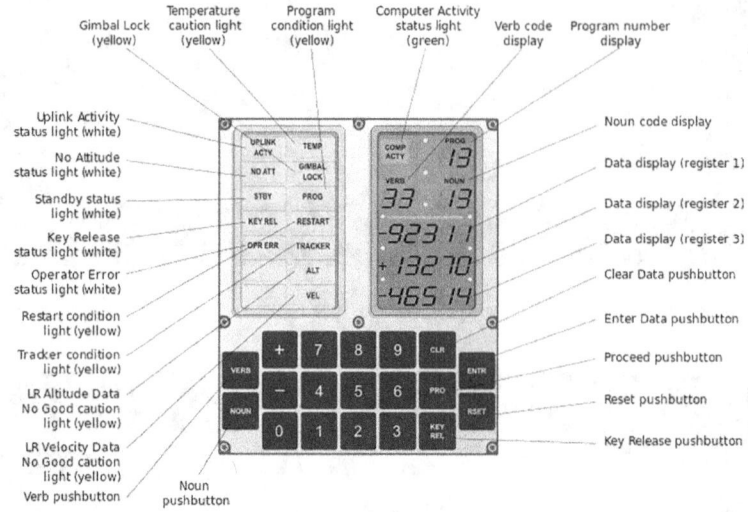

Picture 33 DSKY interface

The control system was completely logical and easy to use (e.g., I want to get from point A to point B). For this procedure, commands with a verb and a noun were used, e.g., LAND (enter this verb) at POINT XXX (enter this noun).

There were 100 different settings between verbs and nouns:

Example:
V37N31E
Verb 37 (change the program) + Noun 31 (which program)
V06N18E
Verb 6 (change the speed)
Noun 84 (speed to xxx)
V50N18E
Verb 50 meaning „change the angle"
Noun 18 meaning „what angle"

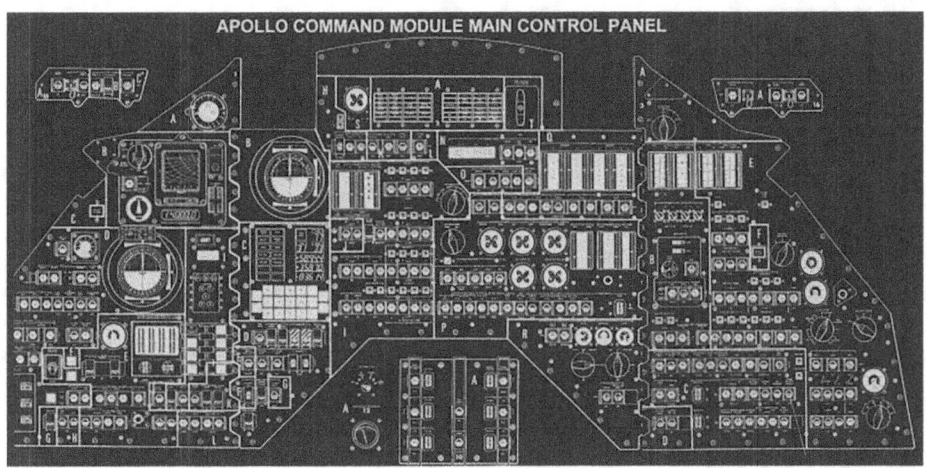

Picture 34 The DSKY position on the command module control panel

3. **The Abort Guidance System (AGS)** has never been used. It would have been used only if other systems had been down, and/or in the case of a failed landing attempt, etc.

AGC structure

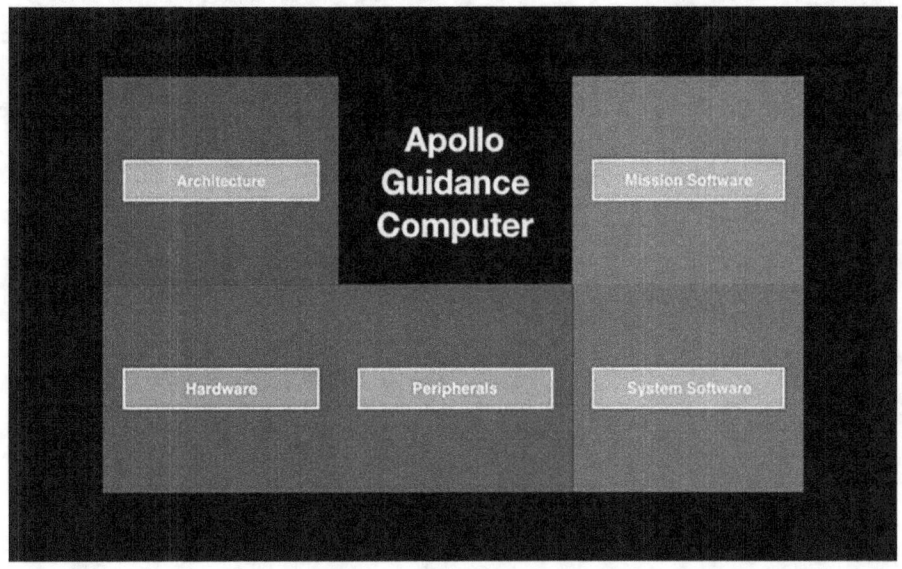

Picture 35 AGC structure

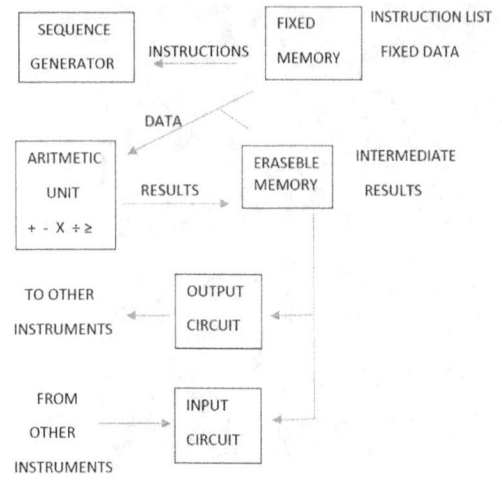

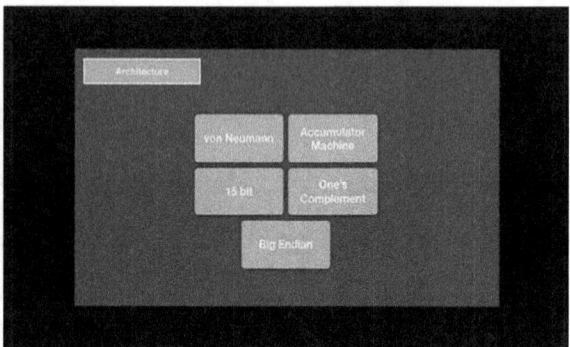

Picture 36 Concept block process diagram   *Picture 37 Fuctional block process diagram*

1. Architecture

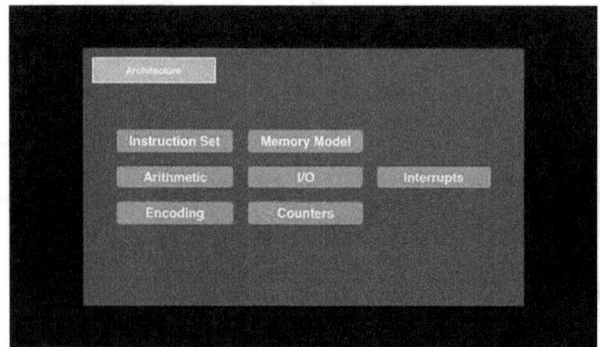

Picture 38 The AGC architecture

For data processing, AGC had 36 instructions:

9X Load Store

10x Aritmetic

1x Logic

5x Control Flow

4x Interrupts

7x I / 0

2. Hardware

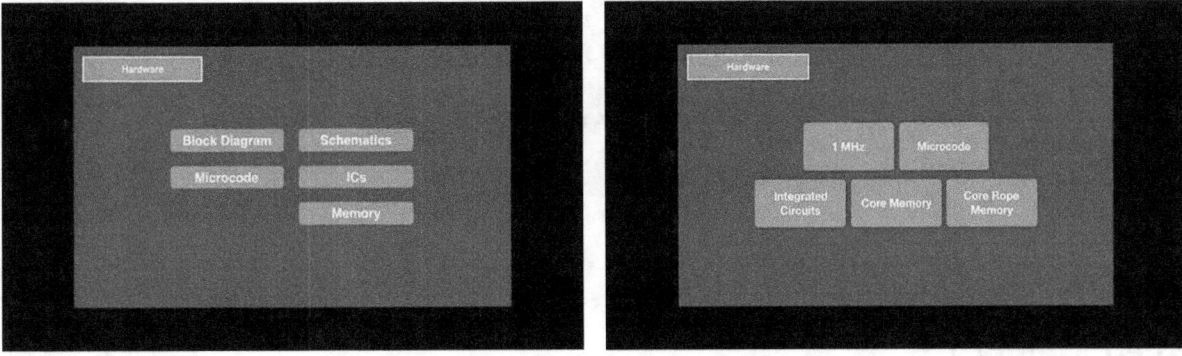

Picture 39 Hardware structure

Memory

In the AGC, there were ROM and RAM sections. The erasable RAM capacity was always available for the most important commands at the given moment. The memory contained 4096 memory cells (1024 Words RAM-magnetic core memory and 3072 Words ROM - string memory) each of which contained:

-15 bit word + 1 parity bit, and it comprised
- numbers between 0000 and 7FFF

The AGC used around 4000 integrated circuits and a unique ROM solution, known as core memory through which a wire was routed. Each IC had a dual 3-input NOR gate. The basic string memory is read-only memory (ROM) for the computers that were used in NASA's automatic Mars probes for the first time in the 1960's. Instead of magnetizing the core in either clockwise or counterclockwise directions by 0 or 1, the core was handled as the discrete wiring transformer core. The wires running through the core stand for 1 and those passing by the core stand for 0. Unlike the conventional magnetic core that used to be used for random access memory (RAM), the ferrite cores in a core rope are used as transformers. The wire signal passing through a given core is connected to the bit line wire and interpreted as a binary en, whereas the line of the word line that goes round the core is not connected to the bit line wire and is read as zero. In the AGC, up to 64 wires could pass through one core. The wires were cast into the water-resistant epoxy plastic.

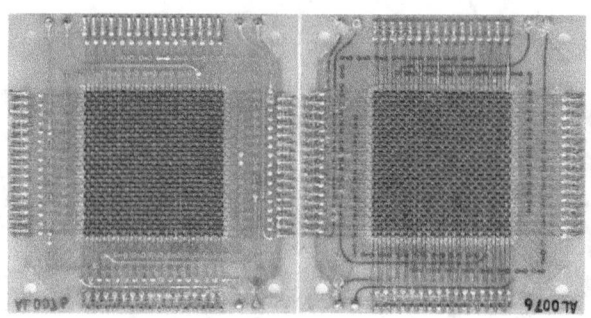

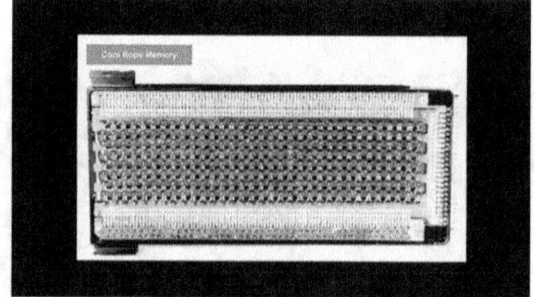

Picture 40 Hard-wired ROM   *Picture 41 Magnetic core RAM*

Banked Memory

The ROM and RAM organization was very innovative. Each of them had Switch RAM and banked RAM space. Memory was divided into core banks, and address could be handled by indicating first which bank was needed, and then the address was given to the bank to get additional available space.

Banked RAM next to fixed RAM is extended to 8 such memories. It depends on the EB 3-bit register command which one will be active.

Banked ROM next to fixed ROM is extended to 32 such memories. It depends on the EB 5-bit register command which one will be active.

The command basics were navigation parameters and the data stored in the program that constituted the whole of the corresponding control device commands. If it was necessary to store data in RAM, they were stored. Otherwise, they were deleted so that memory remained available for other commands. Fixed ROM contained routine commands, constants, coordinates of stars, points and other data that could not be deleted.

ROM was organized in 1024 word banks. The lowest bank (bank 0) was interchangeable memory (RAM). All banks over bank 0 would be fixed memory (ROM). Each AGC command had a 12-bit address field. The lower bit (1-10) addressed the memory inside each bank. Bits 11 and 12 selected the bank; 00 selected the erasable memory bank; 01 selected the lowest bank (bank 1) of fixed memory; 10 selected the next one (bank 2); and 11 selected the BANK register that could be used to select any bank above 2. Banks 1 and 2 were called fixed-fixed memory, because they were always available, regardless of the content of the BANK register. Bank 2 and higher were called fixed transfer, because the bank selected the BANK register.

The address space was expanded by using BANK (fixed) and E-BANK (erasable) registers so that only the current bank account could be the memory of any kind at a given moment, including a small amount of fixed memory and erasable memory. In addition, the BANK register accessed 32K addresses. Therefore, SBANK (superbank) requested access to at least 4k. All interbank subsidy calls had to be triggered from the fixed-fixed memory with special features to restore the original bank during the return. This is basically a remote indicator system.

Thanks to such organization, the available space greatly increased in the most critical situations.

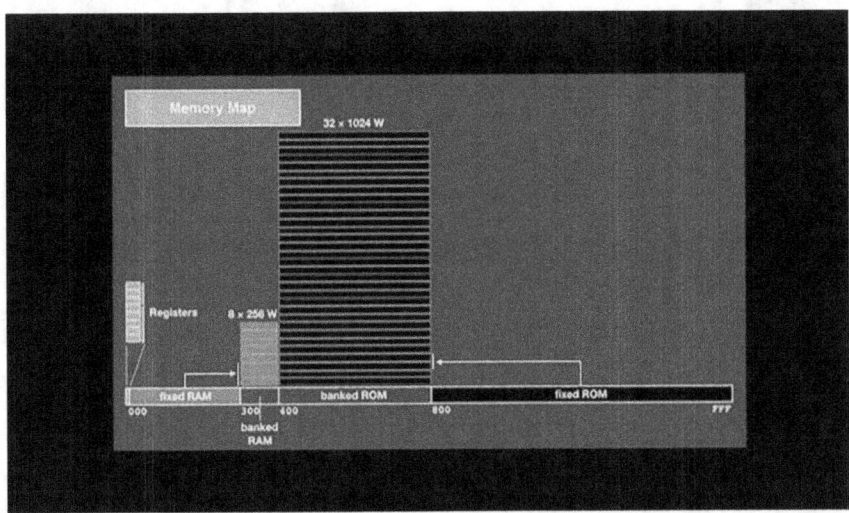

Picture 42 AGC core memory ROM and RAM

Standby mode

AGC had a power-saving mode controlled by the STANDBY ALLOWED switch. This mode turned off the AGC power, except for the 2.048 MHz clock. AGC was in standby mode most of the time.

It was not used to carry out PINC instructions that were needed for updating the AGC clock for real time at 10 ms intervals. For compensation purposes, one function was performed each time when AGC woke from the standby mode and restored real time in 1.28 s. This saved from 5 to 10 KW of energy each time.

3. Peripherals

The peripherals with which the AGC was connected were very advanced. From them, the AGC obtained data and calculated the position of the spacecraft, its angle and velocity, through which the corrections of the path, position and velocity of the spacecraft were automatically determined.

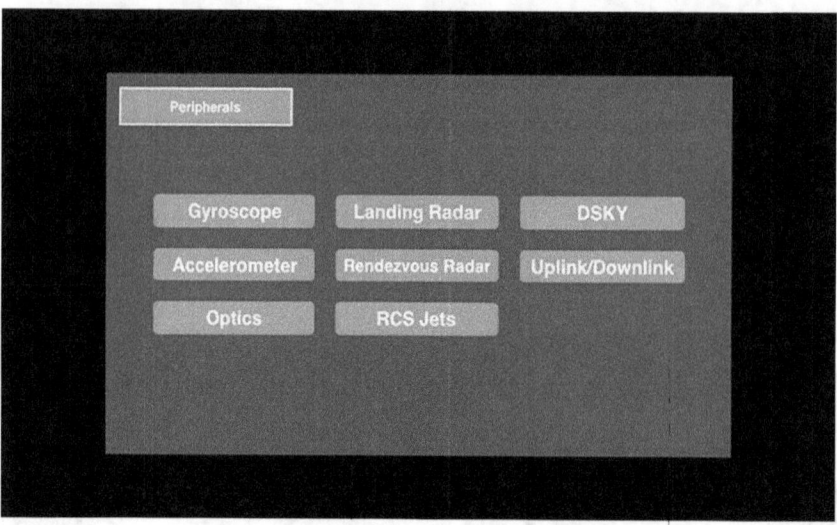

Picture 43 AGC peripheral structure

Gyroscope is a device that works on the principle of keeping rotational quantity. Any new change in turning and moving through three axes is recorded on the gyroscope. Gyroscope knows how much the spacecraft has turned as to the previous situation. Depending on the new position, AGC calculates how and for how long the system needs to turn on thrusters (RCS Jets) to put the spacecraft to the new desired position.

Accelerometer measures three-axis spacecraft acceleration. All three pieces of information go to the AGC that shows velocity and centrifugal force.

Optical telescope (Optics) measures relative positions of spacecraft with respect to the reference points (Earth, stars...). The computer calculates the new telescope position for the new calculation of the spacecraft position.

Landing Radar measures the spacecraft distance to the ground.

The Rendezvous Radar is responsible for the information on the situation of the Lunar module when returning from the Lunar surface to the Command / Service Module (CSM) that is orbiting.

Thrusters (RCS Jets) are small reaction rocket engines installed on the Lunar command and service module. They enable the three-axis spececraft position to move to reach the desired angle of the spacecraft. The AGC determines which of the engines, when, and for how long it will be turned on to reach the new position of the spacecraft.

Uplink / Downlink is radio communication between the spacecraft and flight control

4. System Software

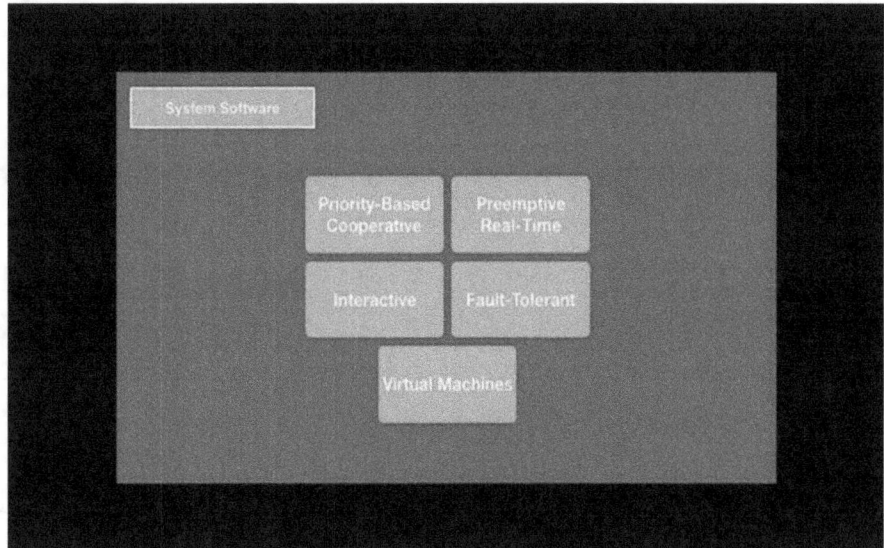

Picture 44 System software

This operating system was called EXEC. It performed priority-based tasks and it could perform up to eight tasks at a time, wherein each task was regularly checked to determine which task had a higher priority. EXEC always had the lowest priority task that needed to be run in order to verify the system state. When this was done, it meant that AGC had no higher priority task at that moment. If it was necessary to store data in RAM, they were stored. Otherwise, they were deleted so that memory remained available for other commands.

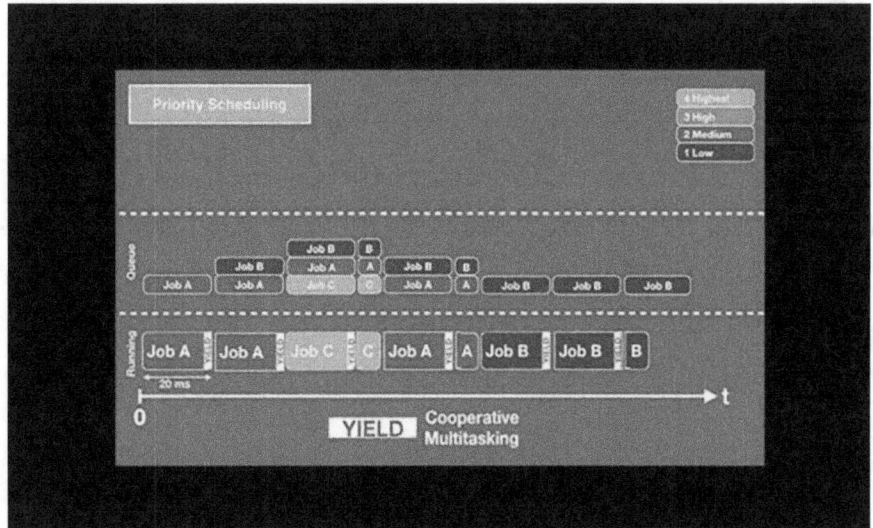

Picture 45 Priority scheduling

Interrupts during AGC operation

AGC had five vectored interrupts:
- Interrupts were triggered at regular intervals to update the user screen (DSKY).
- Erupt was caused due to various types of hardware or alarm errors.
- Keyrupt signaled the button pressed on the user keyboard.
- T3Rrupt was created by the hardware timer at regular intervals to update the AGC real time clock.
- An interrupt occurred whenever a 16-bit word of the data on the upstream connections was loaded in the AGC.

The AGC responded to each interrupt so that it temporarily interrupted the current program, carried out a short interrupt subprogram, and then continued the interrupt program. It also had 20 counters that were memory locations, operating as up/down counters or as exchange registers. The counters would increase, decrease or change the response to internal entries. Increase (Pinc), decrease (Minc), or shift (Shinc) was dealt with a single sequence of microinstructions inserted between any of the two regular instructions.

Interrupts could also be triggered when the counters were overloaded. The T3rupt and Dsrupt interrupts occurred when their counters powered by a 100 Hz machine clock were switched over after a number of Pinc subtitles were carried out. Interrupt cancellation was triggered after the counter that carried out the Shinc sequence moved the 16 bit-connection data to AGC.

5. Mission Software in AGC

Software was divided into the missions, each of which covered an event that was connected to a particular device and was aligned with the task that was intended for this device. Through the I / O interface, the astronaut communicated with the AGC and with a list of commands that were assigned to the device or event.

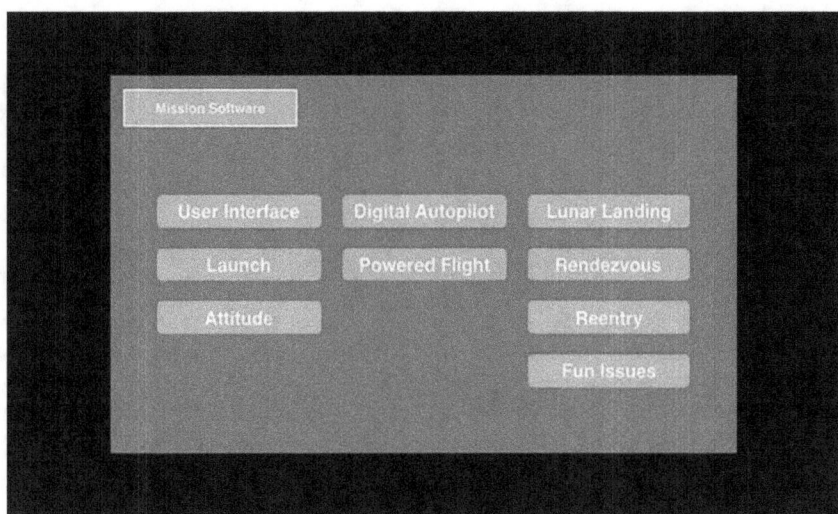

Picture 46 Mission software

User interface indicates that all computer commands were carried out via DSKY, whereby DSKY was also a data viewer. It referred to the Display memory command, Display VERB in memory NOUN. Both commands included two-digit numbers. Astronauts had a list of the most basic commands on the command panel, whereas other commands were in folders.

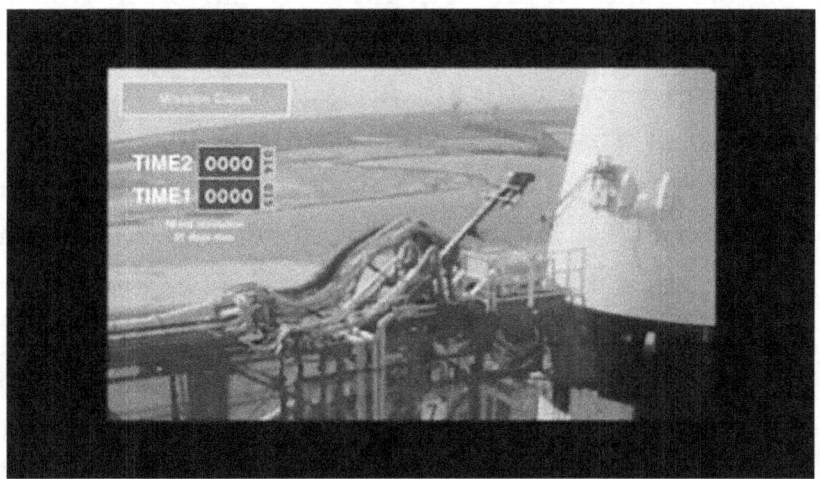

Picture 47 Launch countdown

LAUNCH refers to rocket launch countdown. In the picture below, we see a connector that was disconnected after the launch time started to run. (Zero mission time)

Picture 48 Operations of programs from launch to first stage separation

Altitude and position. After the launch was completed, when the spacecraft was in arth parking orbit and on the path to the Lunar orbit (translunar injection), the spacecraft control was switched over from LVDC to AGC. When they were in earth orbit, they had to be familiar with the spacecraft position. The position was determined by means of a sextant that oriented itself to certain reference points in space. These were most frequently the stars (more than 50) that were in the AGC memory, whereby state vectors were obtained to determine the position and velocity of the three-axis spacecraft.

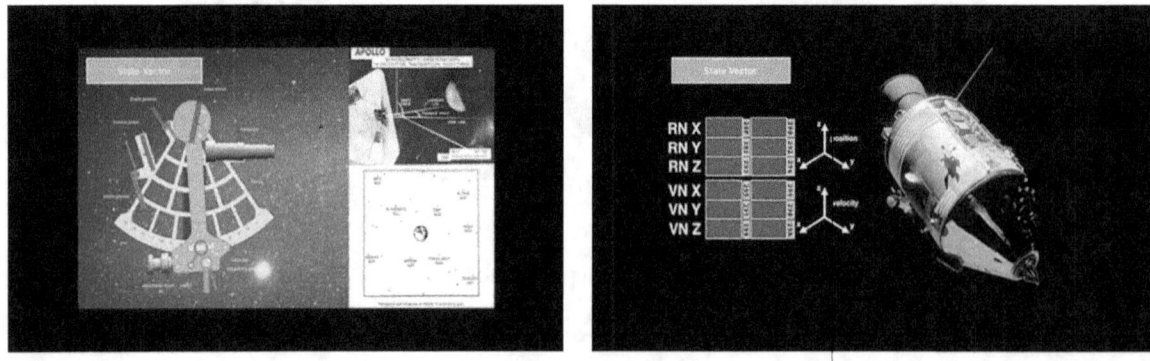

Picture 49 Sextant Picture 50 State vectors

Digital Autopilot

By using a lever, a **digital autopilot** could be used for guiding the spacecraft via electronic pulses. This was a completely innovative solution that was only possible with AGC. It could be used only in the event that the automatic guidance of the spacecraft failed, which, fortunately, did not happen.

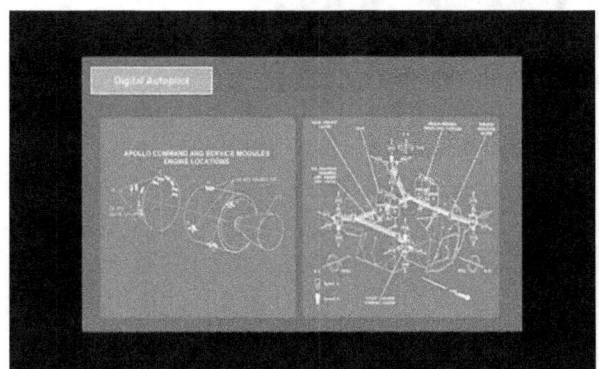 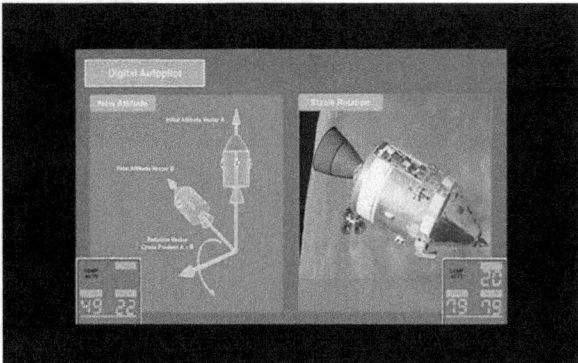

Picture 51 Digital autopilot operation

Picture 52 Ground monitoring of vectors

Powered Flight

When the command/service module pulled out the lunar module from the third stage, the command/service module and lunar module headed for lunar orbit together (trans lunar injection). The launch from Earth orbit to Lunar orbit had to be accurate to seconds. The launch time was determined according to the position of the Earth, the Moon, and the spacecraft position as to the Earth. If this did not happen at a specific moment (time and position), the module could miss the Moon.

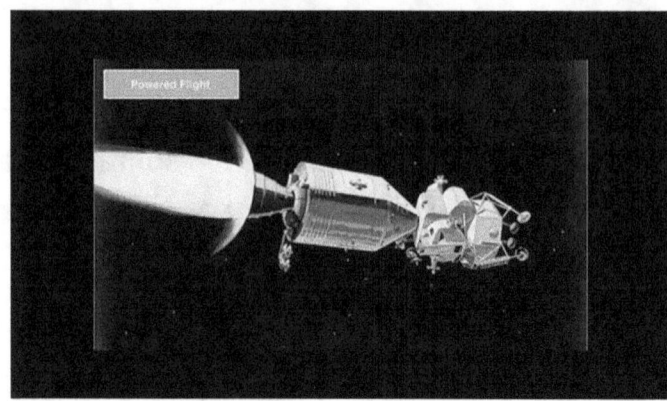

Picture 61 View of the Apollo path from the launch to the Moon and back

Lunar Landing

The Lunar module separated from the command/service module and started braking by using manoeuvering rocket nozzles to enable the module to land in a parabolic curve.

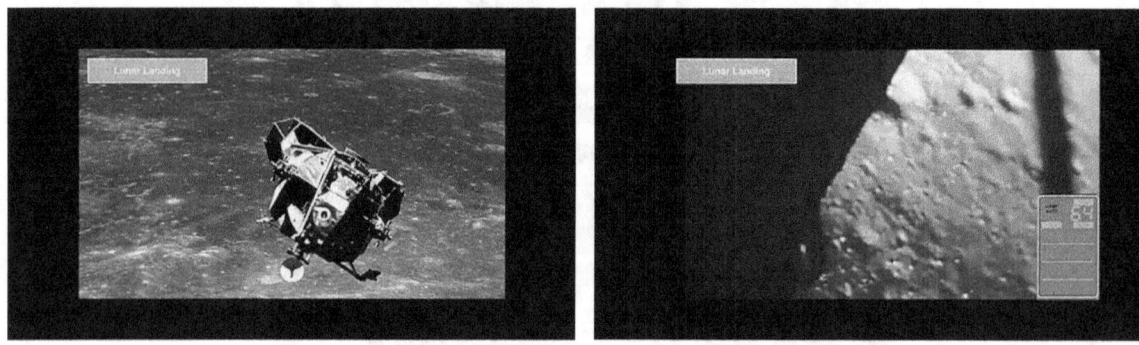

Picture 54 The use of Lunar lander software

Lunar Rendezvous

After the Lunar module took off from the Moon, the landing module remained on the Lunar surface. It served as a launch pad. On the basis of the Rendezvous software, the Lunar module docked with the command/service module which, in the meantime, was orbiting around the Moon.

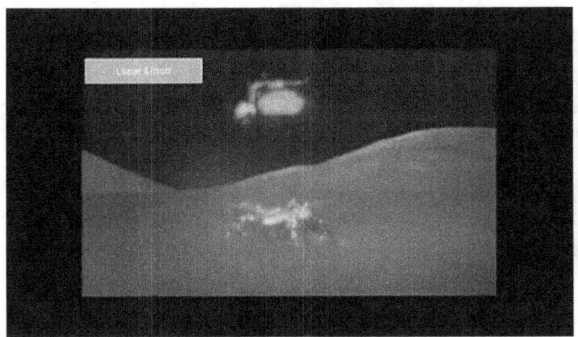 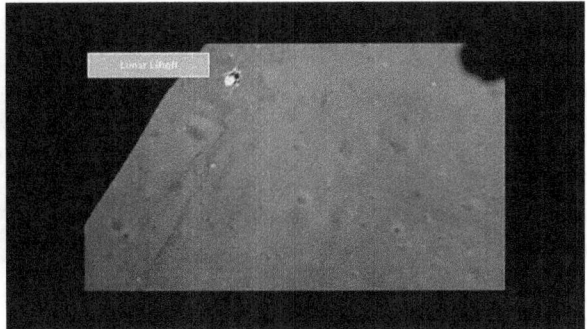

Picture 55 The use of Lunar liftoff software

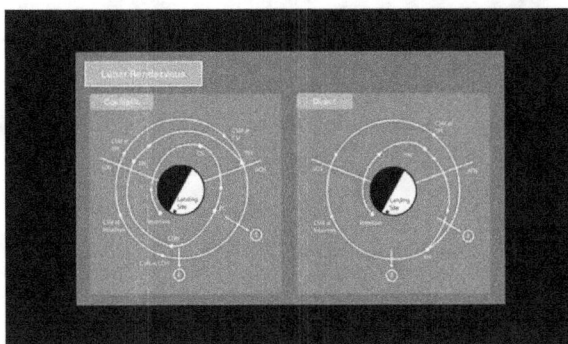 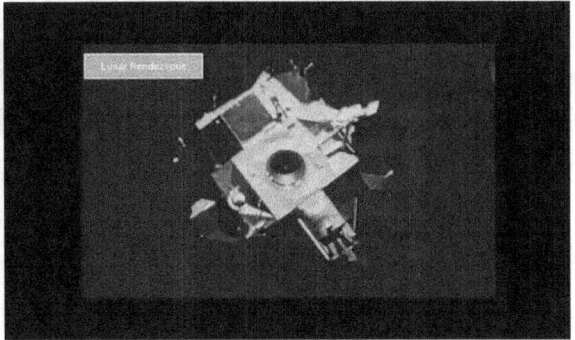

Picture 56 View of the paths for LM-CSM Picture 57 Docking LM and CSM

After docking was done and after astronauts entered the command/service module from the Lunar module, the latter separated from the command/service module and was oriented so that it was on a path to the Sun. Finally, astronauts performed the Trans Earth Injection maneuver.

Picture 58 The Trans Earth Injection maneuver enabled astronauts to fly toward Earth orbit

Re-entry

After re-entering the Earth orbit, the command/service module separated from the rest of the spacecraft and slowed down by means of appropriate software so that the spacecraft was able to enter the Earth's atmosphere at a precise angle. If it were too big, the command module would burn up. But if it were too small, it would bounce off the atmosphere to fly into space.

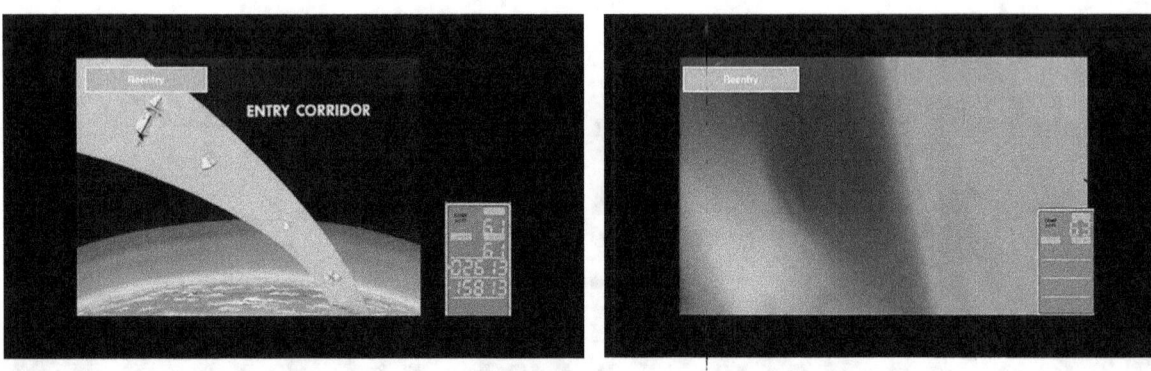

Picture 59 Software for command module re-entry into Earth's atmosphere

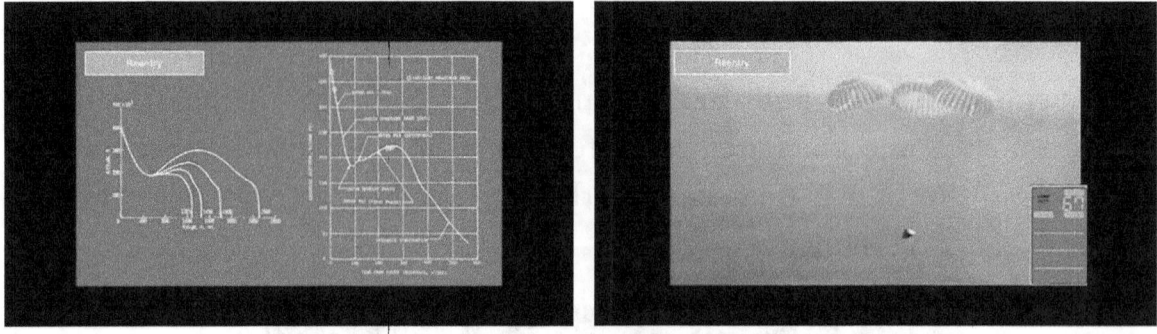

Picture 60 Software for command module landing in the ocean

Mission software by path sections

Each path section had its own software that was precisely determined. All the necessary information was stored in the AGC. This information was used by peripheral instruments and devices for calculating and correcting the path. As seen in the figures above, each part of the path and each spacecraft shift had its own software number. This software took into account the peripheral devices that provided current input data. Thus, the AGC, using the VERB + NOUN command, determined when, how fast and where the spacecraft shift would occur. The computer gave the corresponding command to thrust and brake nozzles in order to enable the spacecraft the desired direction and velocity.

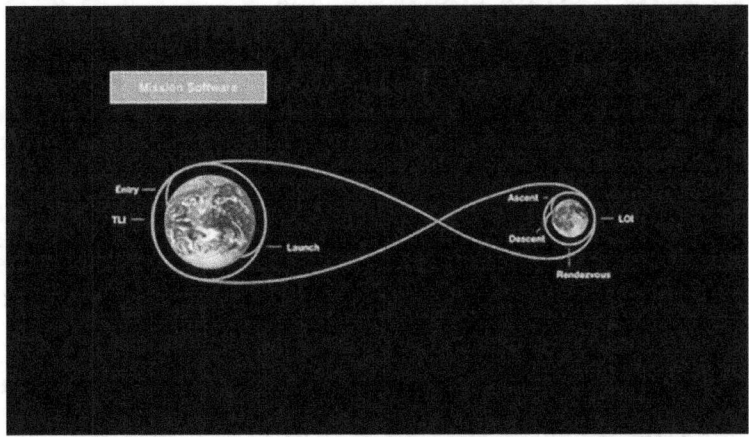

Lunar module	Powered descent
	Landing
	Ascent
	Lunar orbit rendezvous
Command module	Entry
Service module	Trans-lunar injection
	Free return orbit
	Lunar orbit insertion
	Trans Earth Injection
Saturn 5	Launch into orbit

Table 4 Table of path sections that belong to individual spacecrafts

Conclusion

There were no classic operating systems (e.g., Linux, Windows...) that would share tasks between programs, but the Apollo computers had their own operating system (EXEC) that shared more important tasks at a specific time. A more important task was given the entire available space at that time.

Computer had the capabilities of the first personal computer generations 10 years later (Commodore 64, Spectrum, Apple II, etc.). Thus, these computers were 10 years ahead of their time. They were developed as special-purpose computers. Thousands of people worked in the field of computer development at IBM in order to produce computers for performing specific tasks.

RAM and ROM hardware was made manually. The comparison to iPhone, which is 400,000 times faster, is far-fetched because AGC was not a general-purpose computer, but it was developed for specific tasks.

The question is whether the flight to the Moon should rely either on iPhone or on AGC. The Apollo Guidance Computer (AGC) was made to be used in the event of a collision and overload because it was able to reset itself over and over again, thereby continuing the task almost instantly. Today's iPone is not able to do this. This is merely a matter of personal interpretation. The matter can not be looked upon as something that is unambiguous in terms of the processor speed and memory capacity.

It is difficult to explain to the people who have no computer education how it was possible to carry out this mission by using limited capabilities as viewed from the present day perspective. Yes, it was possible to perform this mission by making use of resourcefulness, logic, and ingenuity.

At the beginning of the 60s, the size of the computer with the capability similar to that of the AGC was as big as a room. Prior to the commencement of flights, the computer size had to be reduced to the size of a suitcase. The computer could be used on Earth where there would be enough room for control. Due to a long distance, the problem would arise because a signal would take a considerable amount of time to travel from the command to the module and back, which could mean a delay in the control system. And that would lead to the wrong spacecraft control. Thus, the command module had to be independent of the computer on Earth.

NASA entered into a contract with MIT (Industries Lab.) and with IBM later on. The contract did not specify any specific computer capabilities because nobody knew whether or not it was possible to produce computers with sufficient capability. If this objective had not been achieved, it would have caused the Apollo program failure. There was a fear that the computer might not be sufficiently precise to prevent the command module from entering the wrong orbit. The computer was not yet compatible with the navigation system. Fortunately, this problem was resolved thanks to a sufficiently precise operating computer that was finally produced.

The word "faulty" is displayed in iPhones and modern personal computers if they are overloaded. This means that they need to be restarted, which takes a considerable amount of time at critical moments. AGC performs this task by using Reset Protection so that it can be restarted much faster from on to off and back on again, and thereby resuming the task at the point where it got stuck.

Highly reliable computers were designed for specific tasks. The physical laws of space travel have been known since the beginning of this century. It is about the mathematical formulas that are complicated for non-mathematicians, but not fo rcomputers. The firstcomputers performed up to 4,000 operations per second, whereas high-performance calculators for schools were able to perform much fewer operations. A calculator is essentially a computer for which a person sets up a task to perform an operation. Complex mathematical formulas (e.g., path curve, hyperbola, parabola, etc.) can be produced by any advanced calculator, not to mention the first computer. Of course, the Apollo computers had to take into account a few more variables. In order to determine these variables, a radar, a sextant, a gyroscope, and a telescope were used. The current position of the spacecraft (e.g., distance, velocity, mass, spacecraft angle, endpoint of the path, etc.) was entered in the computer which later determined the thrust and the spaceship angle. Thus, the computer controlled the engines, engine valves, thrust times duration, etc., and it performed sheer calculating operations. There was no iPhone-type computer to control the right path because the computer is used for connectivity, entertainment, and for million operations that are not needed here.

So, the computer performed sheer calculating operations and there was no iPhone-type computer to control the right path because the computer is used for connectivity, entertainment, and for million operations that are not needed here. A lot of space and thereby the computer capability for the current use is intended for snugness and pastime. Therefore, many tasks were left to the astronauts who were the main operating system. They determined when they would use their own expertise and when they would make use of the computer capability. In my opinion, this is an excellent combination. In the past, many modern aviation accidents were caused by a computer that made a wrong decision due to either wrong input data or an insufficient amount of data. Such a task would be better done by man at a certain point of time. This has also been proved by forensic experts, i.e., aircraft accident investigators.

Therefore, I believe that the whole human-assisted electronic system has performed its task more than perfectly.

What does LVDC tell us? If the conspiracy theorists find tody's printed circuit board capable and large enough to launch a rocket, then they must equally feel about the printed circuit board (see in picture) which was as large and capable in 1969 as it is today. It is true that this printed circuit board was developed by many engineers for several years and that this printed circuit board would cost a thousand times more than today's equally capable printed circuit board.

Can a computer be less capable or more capable? I must point out that this has nothing to do with the accuracy of calculations, for example, of paths, speeds, thrust startup time of the second stage of the rocket, etc. The milliseconds are equally accurate in both cases.

The summary of the above-mentioned content is that the commands to the computer were given via the DSKY interface for each path section. AGC did not work automatically all the way from launch to landing in the sea. The mission very wisely and functionally shared work between astronaut and computer. Thus, almost complete human control over the entire flight was achieved.

The Apollo program development means the biggest leap in computer science development in history. If there were no Apollo program, computers would not be at such a stage of development today. Future generations should find out where everything started and where everything came from. It all started with AGC integrated circuits. The Lunar module was the first spacecraft that was controlled by the fly-by-wire flight control system. Today, all modern aircrafts are equipped with this system. By using the Lunar module (LM) to land on the Lunar surface, astronauts considered this the greatest innovation in the Apollo program. In the sixties, the majority of computers took up space in the size of a room, and AGC took up space in the size of a briefcase. This pioneering computer had the first real-time flight data processing system. It also provided mission-critical calculations for the command and Lunar modules. AGC was used for three more Skylab missions, and in 1975, it guided the Apollo spacecraft for docking with the Soviet Soyuz spacecraft. Eldon Hall, lead designer of the AGC, noted that if they had known what they learned later, they probably would have concluded that there was no solution with the technology of the early sixties. AGC was able to simultaneously perform 40 processes, and at the same time, it allowed interrupts of various sensors and entries made by astronauts. It could also receive commands from ground computers, and the mission could also be controlled by means of an optical inertial navigation system. It was able to do anything, including receiving more or too much radar information. In addition, it was robust enough to survive a lightning strike during the Apollo 12 takeoff.

A significant share of all the integrated circuits in the world was designed for the AGC production. At that time, one computer was much powerful, but the AGC was characterized by simplicity and by very easy programming and maintenance. It contained only absolutely necessary software. As Gordon Bell, the father of the minicomputer at DEC, writes that the most reliable components are those that are left out. The AGC was a milestone in hardware and software design and development. It laid down their foundations. The AGC speed, power, and size requirements led the entire subsequent industry to make the first steps in the Moore curve.

The best thing about the mission was that computer never prevailed over man.

Outstanding computer work
- All flight phases were mathematically examined.
- Software was drawn up and a computer was built for these tasks.
- The necessary tasks were software commands entered via DSKY
- It was a purpose-built computer that performed no other tasks or applications. These were sheer work commands.
- No applications, no extra work, no loss; this looked as if we had had an internal combustion engine with no losses. For example: we need much less power for the same work, nothing is lost.
- It was sheer computer-aided flight control without unnecessary data that would slow down or block the AGC.

Still today, we should follow the AGC example and learn from it how simply and logically a computer should work, and how it should communicate with man. Human decisions were still made first.

2.6.7. Apollo 1-10

The Space Agency wanted Apollo to fly for the first time in November 1966, but the problems were accumulating and the day of the flight was moved first to December and then to January 1967. The pressure was high, mainly due to the success of Gemini, or because of the unknown things what the Soviets were doing. At that time, the public did not show a special interest, the costs grew, the congress was concerned. The spaceship was so big and complex that it was difficult to assess its reliability. In addition, a number of mistakes were discovered, supposedly 20,000, for which they hoped would not be essential for safety. Apollo 1 was built in North American Aviation, the other factory than the Mercury and Gemini were built in, so the experience gained from the previous builders of these ships could not be just passed on to Apollo, so these people had to learn everything anew. Plans were changing over and over again because they had problems with the radio and propulsion system, with oxygen supply system and cooling system. The spaceship was so complicated that they were making the system after system and incorporated them into the whole. A total of 300,000 people worked for it in 20000 companies.

Before the Apollo 1 flight, the problems had been accumulating. Controls did not take place according to schedule; they left out an experiment in which a ship should have been charged with 100% oxygen. The first test flight of the first Apollo spacecraft was really intended for the basic test on the journey around the Earth, and for testing the rocket, as well as to the handling and operation of the command module. Next thing was a countdown test, during which all the systems in the ship would have been tested, as with the right countdown. This time, the spaceship was filled with oxygen for the first time, **and soon after this the spacemen entered.** They noticed a lot of loose, intertwined wires around them and they began to complain that the cabin smelled of sour breath. Therefore, the controllers came in to find out the mistake, but the odor disappeared. After that, the cabin began to smell, but the controllers did not find anything unusual after the inspection. They had even more problems, complaining that they poorly heard the control. Nevertheless, the administration approved the start of preparations for the flight. Because the command module was hermetically sealed, they pumped the needed oxygen in the air, even to the point that an overpressure was created. Before the flight and turning on the instruments, there was a spark among the wires which, of course, immediately caused that oxygen would catch fire in the cabin. Despite the prompt call of the astronauts to save them, the rescue workers could not reach them in time. They spent a few minutes to open the door of the command module, which was too late to solve the astronauts. After extinguishing the fire and opening the door, they could only confirm that they lost all three astronauts. What was to be done, they wondered; what went wrong? The experts had to solve the case on their own. There was no external commission, because there is no commission for such a complex system with such knowledge. Even under normal circumstances, no agency was as critical to its own work as NASA. Over the next few months, more than 1500 experts inspected the burned scraps. After ten weeks the committee submitted a report, which was a real masterpiece of self-analysis. They found faults in the electrical installation, an unprotected electric arc and pure oxygen, which had three times greater pressure in the cabin than in the space, so that no extraterrestrial germs would penetrate into the cabin. If this were to happen in the space, oxygen would simply be released into vacuum. The committee made major changes, unprotected wires had to be protected, the flammable substances were replaced, and the opening time of the flap was reduced to seven seconds. In just three months, the space race turned from dazzling success of the Gemini program into disaster with Apollo 1.

Negligence happened because of the race with the Russians. Ironically, three months after the fire, the Soviet Union launched the Soyuz 1 into space, which was maneuvered by Colonel Komarov. During the landing around the circle path around the Earth, parachutes failed to open, and Komarov was dead as the first Russian.

In both countries, they were wondering about the justification of such flights; on the one hand, there was a shortage of money, and on the other the security of the astronauts was questioned. James Webb, the director of the agency, convinced the president to continue working with arguments that America was already far ahead with the project and that it would be a pity if they stopped then. Apollo got a new impetus.

On November 9, after the elimination of all mistakes, the Apollo 4 (Saturn 501) rocket was also standing without a human crew. Apollo 4 - 92 engines, 20 km of wires, high as 36-storey house. Eight or ten seconds after the ignition, when the flames of her mammoth tail were bursting, and thousands of gallons of water tried to control the heat before smelting the pad, it was seating still. Then, when the lockings that held it back were loosened, it left the takeoff pad. It was trembling. Its roar was horrific. It was a wild moment when the engines were swallowing 15 tons of fuel per second and bursting a 250-meter-long flame in the fight against the force that holds the man in the armchair, against the gravity. She grunted and howled. There is no gentle expression to describe its voices. Then it moved. The rocket returned successfully on the orbit around the Earth and mathematical harmony in space. Everything worked fine; the launch was punctual to the second. All three stages of the rocket and all systems went on test. The return was successful; in short, a perfect flight was carried out without people. This gave an impetus to the whole agency and America. Surveyor 1, which landed on the Moon in 1966, contributed to this splendor. More than 11000 photographs were sent to Earth, which were used to study suitable places for landing of the man. The Surveyor program cost 560 million and seven Surveyors, the last landed in January 1968, sent 90000 photos to the Earth.

The Soviets were doing amazing things in the meantime. In 1967 and 1968, they twice launched Kosmos satellites without a crew, approached and docked them without being led by a man.

In January 1968, the United States launched the Apollo 5, an unmanned spaceship, carried by the rocket Saturn 1. For the first time, Apollo 5 also took the lunar unit "spider" with it, which then experienced a baptism by fire. In April of that year, Apollo 6 flew with a Saturn 5 rocket without the human crew, and a number of engine faults were discovered during the flight, but it was possible to quickly eliminate all.

On October 1, 1968, the Space Agency celebrated its tenth birthday. It was preparing for the flight of Apollo 7, which was scheduled for October 11. They reported that in a decade they launched 234 spacecrafts, performed 191 successful flights and 176 fully accomplished tasks, and that the Apollo 4 famous Saturn 501 weighed more than all other American spaceships together. Apollo 7 flight was being prepared and many things had changed for safety. The responsible authorities in North American reinforced the control over the whole process. The head of the spanner that was found in the cabin after the fire became a symbol of negligence and a warning that it was necessary to be more cautious. The special list limited the number of people who were allowed to enter the spaceship. The guard, who stood at the flap, inspected each screw, nut, and a piece of instrument that someone brought to the cabin. All the work was accompanied by television cameras. Women had to remove all cosmetic preparations, and men had to take a shower after haircut. Wally Schirra was the world>s oldest spaceman at his 45 years, who flew into space. He decided that the flight with Apollo 7 would be his last. Schirra>s companions were Walter Cunningham and Donn P. Eisele, the first was pilot with the U.S. Marine Corps, and the other had a doctorate in physics.

Werner Von Braun warned that the Russians may have been preparing a rocket that could take Russian spacemen to the Moon. He was warning because he felt that, at that time, the American public was despondent to the flight into space. On every occasion, flight supporters were emphasizing that it was a race with the Soviet Union, which was accepted with mistrust by some people. President Johnson was clear: Even though the space program will not give us anything else than a photo satellite, this is worth ten times the money that had been invested.

On October 11, the Apollo 7 spacecraft flew into space for an eleven-day flight in which the spaceship behaved very well. Problems emerged with crew due to the overfilled program. Spacemen had to navigate the spaceship with their own engines, to approach to the extinguished stage of the rocket, and besides that, they would still have to transmit all this on television. Shirra did not want to hear about this, but at the end of trying to convince him, he only agreed and pleased the television viewers. The audience watched a seven-minute visit with a great interest; they received a greeting and a saw a pleasant cabin. The spacemen carried out the task and landed safely in the Atlantic Ocean. Apollo 7 was a flawless flight. They achieved some records with it, like the first man to fly three times into space. They transmitted the first TV image from the spaceship, what the Soviets were doing all the time. Apollo 7 was the first American spaceship for a three-member crew, the spaceship the Soviets had already had since the Voyage I. With this flight they carried out all the tasks and the question was how much more a human could do in the space or how much he still could put on a one day of program or in general? What about a flight around the Moon? There was still a rush.

With Apollo 8, they should already have tested the „spider", a lunar module that would take the spacemen to the lunar surface and back, but the "spider" was not ready yet. The excellent success of the Apollo 7 created a great confidence in the

Apollo spacecraft, which was then best used for a safe landing on the Moon. Trust in the technique meant that landing on the Moon could be realized. The Apollo 7 crew met with the Apollo 8 to hand over the experience: to forty-years old Lowell who spent 425 hours in space, Borman who proved himself in the fire investigation in Apollo 1, and Anders, thirty-five years old graduate from United States Naval Academy, and a aircraft engineer of nuclear physics.

On December Sunday in 1968, these spacemen became the first human cargo of the gigantic rocket Saturn 5. The launch facility was standing 5 km away from the rocket for safety reasons. The launch of the rocket was accompanied by unbearable thunder and breathtaking view when ascending the rocket in front of 2000 distinguished guests and thousands of viewers. Saturn 5 was working perfectly, Borman, Lowell and Anders were already in the orbit around the Earth, started the third stage that took them onto the orbit to and around the Moon. Some German newspapers wrote: Today is the greatest day of the century. The crew was offered beautiful images like never before. As they were already far from the Earth, it was already smaller than the window through which they were watching. It shined strongly with the azure and white from the clouds, and the spacemen had problems with flu and vomiting. They noticed that this was happening to all crews in such a way that it was called space sickness. This disease is caused by rapid movements in the zero-gravity space, stress, little sleep, and tense schedules. The next day, Apollo floated between the lines of the Earth's and the Moon's attraction. For the first time, people were directly under the gravitational influence of another celestial body. The next day, they were already 100 km above the Moon and were preparing to ignite the braking engines. Everything worked perfectly and they were already behind the Moon in an area where there is no direct radio connection with Earth. It lasted ten minutes before they were heard again, which was a sign that the spaceship had successfully entered the lunar orbit. In the following hours, images of the lunar surface were being watched on Earth. The back side of the Moon was grayish with full of craters; it looked like a sandbox, like an immense, lonely, and grisly desert, an infinite emptiness. They knew immediately that it was not suitable for landing with a manned crew. They mapped and outlined the places on the front side, where suitable locations for the landing of next crew would be. Nevertheless, the world watched the flight. Dr. Georgi I. Petrov, director of the Soviet Institute of Space Research,

warned that Americans were unnecessarily taking risks but he praised them and wished them a safe return to Earth. During the circulation around the Moon, the world was always awaiting with reserve, when and if the spacemen would respond when they flew from the back side of the Moon. The flight succeeded, the spacemen returned safely to Earth. President Johnson said: "Thank you that we now feel like these Europeans who heard the news about the new world for the first time five centuries ago."

The success of the Apollo 8 mission was also important from the historical point of view, as it rounded up a rather turbulent year both in the USA and in the world. In the year that the war in Vietnam was still raging, Martin Luther King and Robert Kennedy were killed, the streets of USA and Western Europe were flooded by student protesters, and the Eastern bloc was marked by The Prague Spring. However, Apollo 8 returned optimism to humanity. The Time magazine named the crew for personalities of the year 1968. Since they successfully carried out a dangerous and risky mission, for which was also evaluated in Nasa that it had only half the chance of success, this crew had the greatest impact on that year. Commander Borman also received an anonymous telegram: „Thank you, Apollo 8. You rescued the 1968."

Apollo 8 answered some questions. How much work can you give to a person in space, under what pressure they can work effectively, why spacemen feel ill. The Soviet Union was confused, its space program was running and slowing down. Two months before Apollo 8, on October 25, 1968, the Soviets launched the first spaceship of the Soyuz after Komarov's death. The next day, another ship was launched to approach to the first one. It really found the other spaceship, approached to 200 meters, but did not dock with it. It seemed that they had problems in mastering the technique developed by the Americans. The Soviet Union welcomed the flight of Apollo 8, saying that the Soviet program would also benefit from these flights. The same scientist also said that the essential difference between the programs was that the Russians wanted to protect manual systems for operating and navigating spaceships even with completely automatic systems, while the USA left very complicated tasks to spacemen. Another scientist said that the Soviets did not want to put so much emphasis on flight with crew until they utilized the collection of scientific data with satellites. In January 1969, the Soviet Union sent two Soyuz-type spaceships, and this time they docked, but not only that, the spacemen even moved from one spaceship to another.

The last uncertain point was the lunar module, a device that was supposed to carry two spacemen to the Moon and then ascend them to the mother spaceship again. This module had to function impeccably, otherwise the spacemen could stay on the Moon forever. The crew, which was supposed to test the lunar module for the first time, included two members who were veterans of the Gemini flights. Jim McDivitt led the

Gemini 4 during the pioneer American spacewalk by Ed White, and Dave Scott was flying with Neil Armstrong with Gemini 8. However, civilian Russell L. Schweickart has never been in space. During the flight of Apollo 9, he had to step out of the spaceship for the first time. The program determined that the three spacemen would test the landing technique on the Moon near the Earth, where they could be easily rescued and where the risk was also slightly smaller. Although the Apollo 9 was slightly less dramatic, its crew was under great pressure. The last day before the flight, they had also eighteen hours of training. The boys had cold, they were exhausted, which delayed the launch for three days, and that meant half a million dollars. All the spacemen so far somehow got sick, before, during or after landing.

Finally, on March 3, 1969, Apollo 9 was propelled around the Earth. It was flying behind the spacemen in a third-stage section, similar to a garage, the lunar module. When they were in the orbit, the spacemen separated themselves from the third stage, rotated, docked with the «spider» and pulled it out of the «garage». When they were within a safe distance from the third stage, the spacemen observed how they propelled it from the Earth, and the third stage started the journey around the Sun. Then the spacemen docked with the „spider", tried the „twin", and ignited the rocket engines to see if the system was giving anything way during maneuvering. The system was solid as a rock.

On the third day of the flight, McDivitt and Schweickart climbed through the tunnel in the nose of Apollo 9 into the "spider" and tested its main landing engine, which would soften the descent towards the lunar surface. The next day, Witt walked through the space for whole 38 minutes. This walk showed that it was possible to save the spaceman from the lunar module if this did not want to dock with the mother spaceship. The spacemen were split into the lunar module and mother spaceship and separated, and then, with help of their own rocket engines, they reached each of their orbits. Eventually, the spaceships were more than 150 km apart. Then the spacemen tested the spider's landing engine, a rocket with a regulator, which could adjust its power, falling speed. McDivitt dropped the lunar module pad from which the spider would take off from lunar surface and leave it there. Now that they were armed only with the ascent engine, they propelled towards the mother spaceship in order to carry out the approaching and docking, which would have to be repeated 100 km above the lunar surface in four months. First, they were navigating with radar, then they controlled the "spider" with a naked eye. The ships were slowly approaching and, finally, they docked without thrust. Then they finally separated from the lunar module for the last time, and propelled it towards the Sun. Apollo 9 was successful, and now there was an open path to test the lunar module near the Moon.

The Apollo 10 crew could prepare itself based on the Apollo 9 experience. They hoped there would not be any more colds which each of them cost half a million dollars. In the days before the launch in May 1969, the spacemen had been working six hours a day. These were veterans, Young, who flew with Grissom in Gemini 3 and was later also the captain of Gemini 10, Gene Cernan, Stafford's colleague at Gemini 9 who was then also walking around the space, and Stafford who was the best student in his class at the Naval Academy.

The Apollo 10 crew made great efforts to be the first to land on the Moon. The program predicted a cautious flight on which they could gradually complete all of the tests given. The plan was to propel from the orbit around the Earth, and then into the orbit around the Moon, and after that two of the spacemen, Stafford and Cernan, would descend with a spider barely 15 km above the surface, but not to the Moon.

This crew flew into space on May 18 and entered the orbit around the Earth. After the ignition of the third-stage engine, they propelled towards the Moon at a speed of nearly 40,000 km per hour. Observers in Australia saw a spaceship when it was moving like a star. When the engine was ignited, one ninth began to glow, said one Australian man. When the spaceship floated towards the Moon, the Apollo 10 crew turned it around as planned, and pulled out the lunar module from the third stage. Then the rocket was re-ignited from Earth and sent on a circular journey around the Sun. On the journey towards the Moon, the spacemen transmitted television color images. All TV network programs were topsy-turvy, and the recordings they saw were fantastic. The Earth was becoming smaller and smaller and soon they reached the gravitational line between the two celestial bodies and approached the back side of the Moon. The rocket engine had to be ignited at a specific time, so that the spaceship could be orbited into the lunar orbit. The crew described how the Moon was illuminated by the Earth's light from the backside. They also described the colors of the Moon, the description of which differed slightly from the Apollo 8 crew. Critical moments arrived the next day when they had problems with pressure in the binding area and with contact rings. When they arrived behind the Moon, two separate objects, a lunar and command module, appeared from behind the Moon. The lunar module had to ignite the braking rockets due to the change of the orbit, which allowed the spacemen to descend 15 kilometers above the lunar surface. If the rocket had burned for only three seconds too long, it could have crashed into mountain peaks up to 10000 m high. With 5600 km per hour, the module approached the landing site for the next crew. It was reported that there were many rocks in this place with a reflection of different colors. They also tested landing radar, which led them precisely above the landing site, which was smooth with several craters. Then they pulled off the landing pad of the spider and ignited the thrust rocket that propelled it to the command module. Just when the two parts separated it began to bounce, then Cernan shouted to Stafford to hit "AGS" to repair the stabilization system, and spider calmed down. Later, they established that the fault was in the flight program. Both stabilization systems remained turned on, this was a mistake made by someone on Earth. The return later went smoothly. However, the experience they gained during the flight were priceless. The landing site, said Stafford, had 20 to 30 percent of clean land, and if the lunar module had enough time to hover in the air, it would not cause any problems. Would problems with pressure, radar and engines be resolved until the flight of Apollo 11 in July? They got a lot of answers, but there were still many critical issues.

2.6.8. Landing on the Moon 21. 6. 1969

On the launch pad 39A, the Saturn 5 rocket was waiting to make history. It was the Apollo 11 spaceship. It was the spaceship with which people first landed on another celestial body.

Picture 24 Simbolic picture of the spacecraft and its destination

The first Pionir space probes had been already reaching for the Moon, but they did not reach it. This was accomplished by the Soviet Lunniks. They circulated it and crashed onto it, followed by the American Rangers, and then by the Surveyor space stations that landed slightly on the surface. The Orbiters photographed all the Moon's faces, and Apollo 8 watched it from a height of 100 km and saw a black and white emptiness.

Apollo 10 came even closer to just 15 km above the Moon and saw a diverse surface that reminded of the desert and plateaus, of mountains and brownish soil.

Three astronauts were preparing for the Apollo 11 flight, the first being Neil Armstrong who was born on August 5, 1930 in Wakaponeta, Ohio. As a boy, he worked for 40 cents an hour so he could learn how to pilot. At the age of sixteen, he received a pilot's license, even though he had not yet had a driving license. A graduate of Purdue University, Armstrong studied aeronautical engineering, and later also became a fully qualified naval aviator at Pensacola. After graduation, he was improving his knowledege a the University of Southern California. During the Korean War, he was piloting the jet fighters Panthers. He carried out 78 combat flights altogether, and after the war, he entered NASA and flew with an X-15 aircraft up to 70000 meters high at a speed of 6000 km per hour. He was incommunicative, straightforward and had analytic intellect of an engineer. The other one was Edwin E. Aldrin, who was designated as a second man to step onto the Moon. He was born on January 20, 1930 in Montclair, New Jersey. He finsihed The Military Academy at West Point as the third in his class among 475 students. He later obtain his Sc.D. degree from the Massachusetts Institute of Technology. The topic of his doctoral dissertation was the approaching of spaceships during the flight, thus laying the foundations for the Gemini program. He was an Air Force colonel, and later in 1963 he became an austronaut. He also studied geology to understand more the rocks on the Moon. In general, he was known among all the spacemen as the most scientifically oriented. He worked late in the night until his analytic intellect fully solved the problem. The third was Michael Collins who was given the task of flying around the Moon in mother spacehip, while his colleagues would land on the lunar surface. He was born on October 31, 1930 in Rome. Collins graduated from the West Point Military Academy and was Air Force lieutenant colonel. He had flown into the space with the Gemini 10, and walked around the space between the spaceships Gemini and Agena. Initially, he was destignated to fly with Apollo 8, but during training he noticed that his legs were weakening. With several operations, he had the bone excrescence removed in the upper part of the spine. Otherwise, he was relaxed, he had a great sense of humor, and philosophical questions put him in embarrassment.

Initially, the Americans anticipated this event with indifference. When the event was approaching, it occured to them that people were leaving, and they would land and walk on the Moon. Impossible. Less than half of the Americans were opposed to send people to the Moon. However, the process went further, regardless of the opinions.

The personal contact of the spacemen was limited as far as it was possible, and last week, the program of the training was eased so that the spacemen could get some rest.

The launch was scheduled for July 16 at 9.32. That night, the Saturn 5 rocket was filled up with 2 362 500 liters of liqueid hydrogen and oxygen. 700,000 people came to the launch in the district; that is more than usually (230,000.) The ignition of the rocket was carried out without difficulties, meanwhile, the giant was poured with 220,000 liters of water every hour. For almost nine seconds, it was hovering in the air and bathing in flames. Five Saturn engines have collected the power of 92,000 locomotives or 500,000 cars. The takeoff was accompanied by 500 000 000 people all around the world. The launch succeeded to 724 thousandths of a second precisely.

At the same time, the Russians launched the Luna 15, and kept silence about its task, however, they assured that it would not disturb Apollo 11. Apollo 11 left the Earth's orbit, and on the second day of the flight, the spaceship was already 181,000 km far. Then, the speed of the rocket was only 5260 km per hour due to attraction of the Earth. The spacemen were in a good mood, slept for several hours, turned the TV camera to the Earth, and showed it to the earthlings in a beautiful shot, increasingly smaller. The Americans doubted that the Soviet Luna 15 would land on the Moon, pick some rock and return to Earth, simply because it was too small to carry so much fuel with it. It did not even have adequate thermal protection for returning through the Earth's atmosphere.

At the distance from the Earth, the spacemen ignited the rocket engines at a speed of 4800 km per hour and corrected the flight direction. The correction was so precise that this was the only correction along the way. During the rest, when they did not have work, they flew "on the grill" so that the ship turned around three times per hour, thus balancing the temperature.

On the fourth day, July 19, when the spacemen were behind the Moon, they ignited the braking engine and orbited into the lunar orbit. At that time, the earthlings were showed a wonderful sight of the Moon, ridges, plateaus, and mountains. The colors were unusual, from brown, brown-yellow to greyish. The color shade was also affected by the angle of incidence of sunrays. The light reflected off the Earth so strongly you could read a book by the window but the spacemen had another work. After seeing the landing site, the spacemen Armstrong and Collins climbed into Eagle - the lunar module.

After the thirteenth round around the Moon, Collins opened the locking, and the spaceships separated. They were flying together for a while, then Collins ignited the engines and flew forward, giving Eagle space to start descending to the lunar surface during the braking. The radar was constantly reporting the current height during the descent. Although the control computer showed some variations, everything went smoothly and the ground was approaching. At a height of 500 feet above the surface Armstrong noticed that it was approaching the surface scattered with stones, so he switched off the automatic landing device, and the pulse rose to 156 for a moment. He did not reveal himself with anything, immediately grabbed the rudder and searched for smoothly landing place, because otherwise automatic device would have led them straight into the crater. Fifteen meters above the surface, dust clouds were raising around the vehicle. The last few meters the ship literally fell to the ground. Everything was calm and quiet, then Armstrong said: The Eagle landed about six kilometers from the planned landing site. In the space center, they were crowding in front of the receivers to see this historical moment. The spacemen looked through the window and saw a flat surface scattered with small craters, some rocks and a small hill in the distance. After examining all the devices which took them a lot of time, they were waiting for instructions from the Earth. "Whenever you are in the mood", they said from the Earth. Slowly they put on heavy spacesuits, released air from the spaceship and opened the flap at 3.40. Armstrong's 84 kg heavy clothing weighed only 15 kg on the Moon. Slowly he descened down the ladder, so that he would not get stuck somewhere. He touched the lunar surface at 3.56. His first words were: „That is one small step for a man, one giant leap for mankind". He was stepping very carefully, almost dragging his legs behind. He established that the Eagle's feet sank for only a few centimeters, and his barely a few mm into the ground. At 4.42, the spacemen unfolded the American flag and attached its upper edge to the rod above.

They quickly put some rocks into their pockest, so that mankind would have something to show if something went wrong and they had to leave in a hurry. The ground was covered with dust, which was just sticking to the feet. The spacemen collected soil and rock samples, then set up a solar wind gauge, a seismograph and a device that would reflect laser beams back to Earth so that scientists could measure the distance between the Earth and the Moon up to 15 cm accurately. The movement was easier than in vacuum, they could jump with little effort. Aldrin wanted to hammer the rod, but the ground was so strong that he strained himself quiet a lot, and by doing so twisted the rod. When they looked at Eagle, they saw that there was no noticeable crater under it, although the rocket swept dust from the ground. When the Sun rose, it was so glaring that they could hardly see anything, even though they had a strong filter on their helmets. It was time to return, so they put all the stones into a white box and took them into the Eagle. On the twenty-first of July, at 6.11.

they closed the Eagle's flap. After all the work done they rested for another hour and at 18.45 took off from the lunar surface. In the meantime, the Soviet Luna 15 crashed 900 km from the place where they landed. Six hours later they were already sitting in a comfortable mother spaceship next to Collins who was constantly circulating around the Moon. Eleven hours after takeoff, spacemen for two minutes ignited the engine of the command module and sailed off towards the Earth.

The return was calm, they fell into the Pacific Ocean. The spacemen went straight in quarantine on the aircraft carrier, so President Nixon could greet them only through the window of the quarantine cabin. He said America and the world were proud of them and invited them to lunch with him when they could leave the cabin. It was a historic event for the whole world.

2.6.9. Apollo 12-17

Apollo 12

The launch of the Saturn 5 rocket on November 19, 1968 was the second manned expedition to the lunar surface.

Picture 25 Insignia of Apollo 12   *Picture 26 Apollo 12 Crew*

Command and service module (CSM - Yankee Clipper)
Lunar module (LM – Intrepid)
Crew: Charles Pete Conrad, Alan Bean, Richard F. Gordon
Duration: 10 days

For the landing site, the crater Surveyor near the Oceanus Procellarum was selected, which was determined by means of the images of the automatic probe Surveyor and the lunar orbiter. The mission of Apollo 12 was, among other things, the return of parts of the automatic Surveyor III spacecraft that landed on the Moon in 1967. The Oceanus Procellarum landing site, which was located 1500 km from the Mare Tranquillitatis (Sea of Tranquility), is especially interesting for geologists because it is flat and covered with material from the nearby craters Copernicus and Surveyor.

Picture 27 Lunar module landing

The purpose was to study the parts of Surveyor III that had been exposed to space conditions, radiation and extreme temperatures for more than two years. Data on the behavior of the material that was exposed to these conditions gave useful information for later possible placements of bases on the Moon. When studying Surveyor robotic spacecraft, they noticed that after three years, the metal changed color from white to dark brown, all due to the sun's influence. They also carried out a scientific package of experiments- Apollo Lunar Surface Experiments Package - (ALSEP). The entire package weighed 112 kg on Earth, and only 18 kg on the Moon, so they did not have any problems with carrying and disassembling. As in Apollo 11, they also had a TV camera and a seismograph rather than a laser reflector. The better were the analyzer of the solar wind and many other physical instruments. For the first time, they set up an electric generator that produced an electric current with the process of plutonium decay and had power of 75 W which was enough to send data from instruments directly to Earth. Heavy equipped, the astronauts traveled a distance of 1300 m at a speed of 4 km/hour. Due to the exertion, their heart beated 90 to 130 beats per minute. In such a case, the control from Houston indicated they had to return to the LM module and rest. They also carried out geological analyzes of the surface of the craters. They studied places on the rim of the craters, where different colors of the soil were detected, from gray, brown to black. The astronauts performed all the tasks, except of the fact that color TV camera

Picture 28 Equipment setting

did not want to work. Their coordination on the surface was already better than with the Apollo 11 crew, but there was a downside due to exertion in overcoming larger distances. They collected 34 kg of rocks and stayed on the surface for a total of 31 hours. The mission and the flight were completely successful.

Charles Pete Conrad led the second mission as commander of Apollo 12. He was the third man to step onto the Moon, and before that, he had been gaining experience with the flights of the Gemini 5 and 11 spaceships. As test pilot for the US Navy, he joined NASA in 1962, and retired in 1973, after being the commander of the Skylab 2 Space Station. After this career, he was working at the American Television Company (ATC) and McDonnell Douglas. He died in 1999 after a motorcycle accident in Ojai, California.

Alan Bean was pilot of the lunar module and this was his first space flight. He was walking more than one day with Pete Conrad on the Moon. Prior to 1963, when he joined NASA, he was the captain of the US Navy. After Apollo 12, he was commander of the Skylab 3 mission in year 1973. He retired in 1981 and began painting, wherein he could express his inspiration from space experiences. In 2009, during the celebration of the 40th anniversary of the first landing on the Moon, he exhibited his works at the Smithsonian National Museum. Bean died last year, on May 26, 2018, after serious illness at the age of 86.

Richard F. Gordon was commander of the Apollo 12 command module. During the circling around the Moon, he was painting a surface where the following Apollo missions could have landed, while Conrad and Bean were walking on the Moon. Before then, Gordon had been a captain in the US Navy when he was selected as astronaut in 1963, and later as pilot of Gemini 11 crew in September 1966. He retired in 1972. He continued his work at the professional football club, Energy Developers Limited (EDL), and at Resolution Engineering and Development Company (REDCO). Gordon died on November 6, 2017.

Apollo 14

Due to the accident and return of the Apollo 13 crew, this was the third manned expedition to the Moon.

Command and service module (CSM - Kitty Hawk)
Lunar module (LM – Antares)
Crew: Alan B. Shepard, Stuart A. Roosa, Edgar D. Mitchell
Duration: 9 days

Picture 29 Insignia of Apollo 14

Picture 30 Apollo 14 Crew

The Apollo 13 accident caused only a four-month delay, so they already took off on January 31, 1971. After four days of flight, the astronauts landed in Fra Mauro. During the landing, Alan B. Sheppard, Stuart A. Roosa and Edgar D. Mitchell observed the Cone crater. At the planned landing site, they encountered a lot of small craters, however, Alan B. Shepard, the commander, found a landing site in the size of LM inclined by eight degrees, which later proved to be the cause of uncomfortable sleeping. In LM, they took some pictures of the outer landscape

Picture 31 Apollo 14 LM

that they sent to Earth, and they looked like a snowy landscape in the night on Earth. After the preparations in the LM and the exit, the astronauts had two walks in total duration of 9 hours and 25 minutes. They collected 42 kg of rocks and first used a hand-pulled mobile vehicle with which they could carry instruments and tools. It

Picture 32 Rocca by LM

is interesting to note that the third stage of Saturn V Apollo 14 landed on the Moon, and the stages on other missions were directed towards the Sun.

After the exit (EVA- extravehicular activity), they set an operating camera with which they could record themselves when they were setting up ALSEP. When carrying tools and instruments, they needed much less effort because of the pull-cart that Apollo 12 did not have.

Moonwalk lasted slightly longer than in the Apollo 12 mission, as they had a rather uneven terrain covered with smaller craters. With the second exit, they wanted to reach the Cone crater which was 650 meters away. They established the path that led to it was anything but easy. During climbing on the rim of the crater (inclination of 10%), they had to make a lot of stops because their heart rate was up to 140 beats per minute. They were told fromHouston they needed to take more time for resting. The last few meters, the ascent was quite steep, and due to the difficult terrain time was not on their side, given the estimated use of oxygen. When they reached the top, they did not actually come to the rim of the crater. It was about a hundred meters away, but now they had to take into account the time, although they were only 30 meters from the rim of the crater. They quickly took samples of rocks and went back to the lunar module. The return was somewhat easier even though they had problems with balance. After a lot of exertion, they were very tired and returned to LM with their equipment and 42 kg of rocks. Apollo 14 was an important mission because the astronauts studied, among other things, the possibilities of using the lunar rover, which was destined for the next expedition of Apollo 15.

Alan Shepard was commander of the Apollo 14 spaceship. Before commanding the Apollo 14 in 1971, he had become the first American to reach the space with the Mercury program in 1961. Apollo 14 was his second flight into space. Before that, he was chief of NASA's Astronaut Office and he continued this mission after Apollo 14 mission. During a walk across the lunar surface he hit two golf balls and watched how far the balls flew at low lunar gravity. Shepard left NASA and the Navy in 1974 as the last US Navy Admiral. After NASA, he wrote several books about his experiences, about the movement on the Moon, and helped to lead the Astronaut Scholarship Foundation for science. He died in 1998 from leukemia.

Edgar Mitchell was pilot of the lunar module, which was his first flight into space and lasted 216 hours and 42 minutes. After Apollo's mission, he retired as a naval captain in 1972 and founded the Institute of "Noetic "Sciences for the purpose of consciousness research. He wrote books about mystical experiences and psychological research. He died on February 4, 2016.

Stuart Roosa was pilot of the command module. On his first and only journey into space, Roosa was painting and observing the lunar surface during the circulation around the Moon while the colleagues were walking across the Moon. After leaving the US Air Force, he joined NASA in 1966. After Apollo 14, he worked on Space Shuttle program. After retirement in 1976, he worked for the U.S. Industries for a commercial real estate company, and was president and owner of the Gulf Coast Coors. Roosa died in 1994 due to complications of pancreatitis.

Apollo 15

Command and service module: (CSM - Endeavour)
Lunar module: (LM - Falcon)
Crew: David R. Scott, Alfred M. Worden, James B. Irwin
Duration: 12 days 17 hours

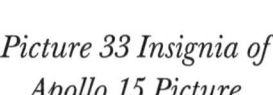

*Picture 33 Insignia of
Apollo 15 Picture*

*Picture 34 Apollo 15 Crew
with Lunar Rover*

The first three Apollo spacecrafts with human crew were breaking new ground on the Moon. The astronauts had to adapt to the one sixth of gravity, walking and testing of the stiffness of the spacesuit during movement, etc. In each mission, they had certain tasks for scientific experiments, which also meant working with tools. The astronauts had to carry some staff, which was not too convenient because, due to stress and stiff suit, every move meant a great exertion, despite the one sixth of gravity. They could not spend too much time on this, because they were limited by the use of oxygen.

This mission was a turning point in journeys to the Moon. The lunar module was so perfected that they could afford extra weight and then take the lunar roving vehicles (LVR) with them to the Moon. With this vehicle, they overcame far greater distances across the surface, carried more equipment with them and thus performed more experiments. Apollo 14 was by far the most advanced mission with the original Apollo technology.

Werner Von Braun had already early imagined traveling on the Moon with three modules. They would have had two, each with twenty astronauts, and the third one would have been loaded with up to 300 tons of cargo in order to have stayed on the Moon for six weeks. NASA preferred to preserve what was financially possible in reasonable time period under given circumstances. The length of stay on the Moon depended on the food, water, and oxygen available. The peak of the early exertions and ideas for the lunar vehicle was a mobile laboratory, as a two-tone two-seater with a closed cabin and with a range of 100 km. But how to get it into the lunar module? LVR (Lunar Vehicle Rover) was the right solution for increasing the productivity. Several prototypes were tested in the desert, and the astronauts practiced its control even in a giant aircraft, in which they achieved short-term weightlessness. His development cost 40 million dollars. Success exceeded expectations, since the astronauts did not experience serious accidents on the Moon, and Gene Cernan reached a record speed of 18 kilometers per hour, which was ten kilometers per hour more than the average speed. The LVR was 3.1 meters long and due to its aluminum construction weighing only 210 kilograms. It was driven by four electric engines with a 0.25 "horsepower" on each wheel. All wheels could be turned around, while two 36-volt batteries with a capacity of 121 Ah provided the electricity supply. The driver was controlling the speed and direction with a joystick. Among the equipment there was also a special navigation system, which recorded changes in direction according to the position of the lunar module and led the astronaut, along the same path, back to the starting point. The vehicle also had a TV camera that could be operated from the Earth.

Picture 35 First Lunar Rover drive

Picture 36 Collecting samples of lunar soil

The position of Apollo 15 Hadley landing was well selected from a geological point of view. There were a lot of structures for exploration, and a landing site was also appropriate. Two hours after landing, the astronaut already put on spacesuit and took the first steps outside, followed by the first drive with a rover in general, which immediately showed a great deal of functionality. Without any particularities, the astronaut traveled the distance of 6 km. The vehicle behaved nicely at 10 km/h, but it was a little bouncing due to the undulating terrain. In such a drive, they praised the indispensable safety belts. In the first geological researches, they arrived to the St. George crater, where they were interested in how it was formed, whether it was of volcanic origin or it was formed by a meteorite fall. They collected rocks of anorthosite and breccias, the composition of which tells much about the origin of formations. Due to the good navigation system, they found LM with no problem when returning, in which they developed the scientific equipment ALSP. Later they were drilling into the surface, which turned out was not the easiest task, because of the solid surface and poor drill. After six hours and a rest in LM, the astronaut went on a second drive towards the south where they noticed rocks with a green shade. As it was later shown by the geological structure, the green rock was iron-magnesium glass. When they examined the other rocks, they saw the green rock again, and found that it had the crystal wall of the mineral plagioclase, different from the breccias and the sea basalt they had been collecting until then.

After it was known that America won the battle for the Moon, the public was interested in the history of the Moon. The crew of Apollo 11 and 12 brought the samples of the lava to the Earth that made the "seas", and the Apollo 14 crew samples of breccias that were ejected from the large pools of "rain." It s called as such because the traces of the ejected material from the Earth can be seen as rain would make them.

They needed to find another missing rock to fully identify the history of the Moon formation, of the so called „The oldest rock". They found a rock anorthite which was estimated at four billion years old and was an important part of the structure. Four hours after the "trip" they returned to LM with help of navigation and with 38 kg of rocks. Before they entered the LM, they had drilled a hole into which they put a device for measuring heat flux. The first two exits, each for four hours, took the astronauts a lot of time. The rest of the time, they should have carried out three walks or three drives. However, as there was a shortage of time, they had to change the schedule according to Houston instructions and to shorten or eliminate the planned walks that included soil research. They also performed deep drilling to take samples of soil. With a great exertion they drilled into the ground about one meter and pulled out a tube with a sample. It was the deepest sample taken so far. Despite the lack of time, they went 2 kilometers with the vehicle to the crater on the other side, and from there they took samples of soft soil. The Apollo 15 crew established that they could do a lot more tasks with the vehicle, and that they could work without a problem for three days, of course, with rest periods. They gained a lot of experience for the following Apollo 16 and 17 missions. They left the surface with exceptional photos and docked with the command module. All three pilots traveled back to Earth with the pilot of command module.

David Scott was commander of Apollo 15. During the fourth mission on the Moon, Scott and Irwin became the first men to drive across the lunar surface with a vehicle. Scott was already a space veteran, since he had experience from the Gemini 8 flights in 1966 and with Neil Armstrong in the Apollo 9 mission in 1969. After joining the NASA in 1963, Scott collected 546 hours and 54 minutes of stay in space. After Apollo 15 mission, he was the director of the NASA Flight Research Center in Edwards, California. He is now a retired Colonel in US Air Force.

James Irwin, was pilot of the Apollo 15 lunar module. This was his first and only flight where he was walking across the Moon around the Apennine Mountains and collecting 171 lbs (77.5 kg) of lunar rocks with Scott. Irwin, as retired General of American Air Force, joined NASA in 1966 and left it after Apollo 15 flight in 1972. Later he became a preacher and founded a religious foundation called the High Flight Foundation. He died in 1991 due to a heart attack.

Alfred Worden was command module pilot. As a test pilot, he was one of the astronauts selected by NASA in 1966. He worked as a scientist at the NASA Ames Research Center in California. He left NASA in 1975 and continued his work at Maris Worden Aerospace and BG Goodrich Aerospace.

Apollo 16

Command and service module (CSM - Casper)
Lunar module (LM - Orion)
Crew: John W. Young, Thomas K. Mattingly, Charles M. Duke
Date of launch: April 16, 1972

Picture 37 Insignia of Apollo 16   *Picture 38 Apollo 16 Crew*

The crew had the same tasks as the Apollo 15 crew. The Descartes Crater was chosen as the landing site. The lunar module was disconnected from the command module at a height of 16 km, and descent went smoothly, although oscillations occurred in thrust motors in the meantime. They landed only 270 meters from the planned landing on the plateau of the Crater Descartes. The LVR and the scientific equipment of ALSEP, which contained instruments for measuring heat flux, solar wind, magnetometer etc., were unloaded. After preparation and exit from LM, they drove to the crater Buster, where anorthite (white crystal material) was found, and at the same time, they established that the Descartes plateau was not of volcanic origin.

Picture 39 Journey with Lunar Rover

They also drove to the Flag and Spook craters, collecting samples from the large wall they encountered along the way. When returning to LM, they set up equipment for measuring solar wind. The first exit lasted 7 hours in which they drove 4.2 km.

Picture 41 Duke's familly picture on the Moon

Picture 42 Collecting sample from the untouched soil

The next exit, they visited various geological points on the untouched Moon's ground. This second exit lasted 7 hours, in which they made 11.1 km. The third exit they went to the deep North Ray crater, where they took samples of soil. While returning to the LM they encountered a Shadow rock wall, a part of which is always in the shadow. This part has never been exposed to the solar wind, so it was determined how this affected its structure. A hundred meters before LM they set a camera on the rover, which was guided from Earth, and accompanied the LM takeoff. The third exit they spent five hours and 40 minutes outside the vehicle, and in total, they collected 92 kg of rocks. Before returning to LM, Duke left a picture of his family on the lu lunar surface.

Before they docked with the command module and left the lunar orbit, another spectrometer had been launched into the orbit. During the journey to the Earth, for the forthcoming Skylab mission, the commander of the command module left the spacecraft, took the film from the instrumental part of the spacecraft, visually analyzed the equipment and performed other experiments. The journey back home went according to plans, and the mission succeeded although it was shortened for one day.

John W. Young was commander of Apollo 16, who led the fifth human expedition to the Moon in April 1972. He and Charlie Duke explored the lunar surface in Descartes and collected 200 lbs. (90.7 kg) of rocks, and drove more than 19 miles (24.75 km) with the lunar rover. Young joined NASA in 1962 as pilot of the US Navy, and had already been six times in space with Apollo. He flew with Gemini 3, 10 and with Apollo 10 with which he circled the Moon in 1969. After Apollo, he flew twice with Space Shuttle, so he logged 835 hours flying time in spacecraft across the space. After his flights, he stayed with NASA until his retirement in 2004. He died on January 5, 2018.

Charles Duke was pilot of the lunar module. As a retired general in the US Air Force, he joined NASA in 1966 and had one space flight with Apollo 16 before retiring from NASA in 1975. During the Apollo 16 flight Duke and Young developed a cosmic ray detector and UV radiation camera on the lunar surface. After NASA, Duke carried out various business opportunities and founded the company Duke Investments Charlie Duke Enterprises.

Thomas "Ken" Mattingly was pilot of the command module. He joined NASA in 1966. He had been scheduled to complete the first flight with Apollo 13, but he was removed 72 hours prior to the launch due to exposure to so-called „German measles". Later, he was assigned to the mission of Apollo 16. During the circulation around the Moon he photographed and mapped the geochemical structure of the surface on the lunar equator. After the Apollo 16, Mattingly flew also with Space Shuttle STS-4 and STS-51 C before retiring in 1985.

Apollo 17

Command and service module (CSM - America)
Lunar module (LM - Challenger)
Crew: Eugene A. Cernan, Roland E. Evans, Harrison H. Schmitt, James B. Irwin
Duration: 12 days 14 hours

Picture 43 Insignia of Apollo 17

Picture 44 Apollo17 Crew

After a successful takeoff and leaving the Earth's orbit, the crew carried out numerous experiments during the flight towards the Moon. Measurement of temperature, lightning from the Sun, etc... Before entering the lunar orbit, a module with scientific instruments had been launched into its orbit. They landed on the lunar surface on December 11, 1972, but missed the landing site for only 200 meters. Four hours after landing, the astronauts already left the LM, planted the flag, and prepared the rover as well as the scientific station ALSEP 185 meters away from LM. The upgraded package ALSEP performed even more measurements than on previous missions. It measured the heat of the lunar surface, performed tests with meteorites, measured seismic activity and cosmic radiation.

Picture 45 Astronaut on the lunar surface

The crew drove with the LVR to the Steno crater, where it took samples and encountered the remains of the lava. They found regolith and collected 14 kg of samples within 7 hours and 37 minutes. The second exit, the astronaut first made scientific experiments with ALSEP, an experiment with electric current in the regolith,

to determine the presence of water, and then drove to 7 km distant crater Nansen. There they found slightly lighter materials than on other surfaces (magnesium silicate) and breccias (young sedimentary walls). When returning to LM, they stopped in the Lincoln Scarp area in which they performed gravimetric measurements, collected samples, and continued to the rim of the Shorty crater. Here they encountered an orange and black surface, which was later analyzed by scientists on Earth and established to be of volcanic origin.

The third flight lasted 7 hours 37 minutes; they carried out seven measurements with a gravimeter and collected 34 kg of rocks. After 15 and half hours of rest in LM, the crew left the vehicle for the third time for another 7 hours and 15 minutes. In addition to the usual works, the astronaut went

Picture 47 Samples with the orange soil

Picture 48 Astronaut at the large rock

north to the Taurus-Littrow valley, where they observed the huge wall on where the breccia and glass basalt were located, representing the remains of volcanic lava, and they collected 62 kg of rocks. In total, they spent 110 hours on the surface and collected 110 kg of rocks and installed 16 automatic measuring station During the circulation around the Moon, the command module carried out numerous measurements, such as measuring the density of the Moon with an UV spectrometer, thermal characteristics, measuring with infrared radiometer, measuring the ground with lunar sounding, and taking panoramic photographs of the lunar surface. Before entering the LM, the crew had left the rover with a camera 150 m away due to the recording of LM takeoff. After docking the LM with the command module, the spaceships flew towards the Earth. During the flight, the commander came out of the spaceship and took the tape from the outside camera. Data from various measurements were recorded on this tape. The return to Earth happened west of the Hawaiian Islands, 1.6 km away from the intended landing. This mission was longer than any previous missions, the most ambitious and productive one. The only scientist Schmitt also stepped onto the Moon.

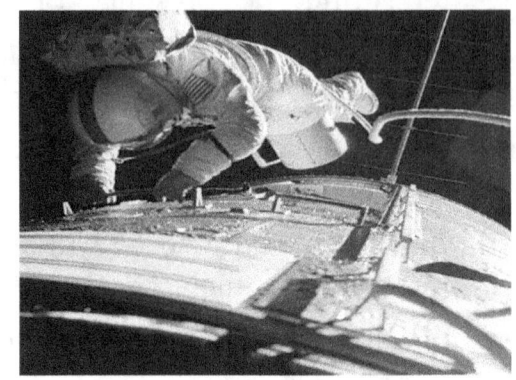

Picture 49 Commander took the tape from outside of CSM

Eugene Cernan was commander of the Apollo 17, and before that, he had been on two flights into space on the Gemini 9 in 1966 and Apollo 10 in 1969. In December 1973 he led the sixth and last expedition to the Moon with Apollo 17. Cernan was the last who left the footprints on the lunar surface. As captain at the US Navy, he retired from NASA and the army in 1975. Later he was employed by Coral Petroleum, and then he founded his own consulting company. He was also the head of Johnson Engineering Corporation, which helped NASA to develop trainings and space exploration equipment. He died on January 6, 2017.

Harrison Schmitt was pilot of the lunar module. He is educated geologist and the only astronaut - scientist who, without military experience, walked across the Moon and helped to teach other astronauts in geological researches on the Moon. Apollo 17 was his only journey to space. After returning from space, he stayed with NASA as an astronaut - scientist, he was NASA employee, organizing NASA's Energy Program Office. In 1975, he left NASA because he was running for a senator in state New Mexico. He won the election and became a Republican senator.

Ronald Evans was command module pilot on Apollo 17. During the circulation around the Moon, he installed a camera and three recording tapes outside the spacecraft. He is known as the astronaut who remained much longer in the lunar orbit than everyone else. He became astronaut in 1966, and in 1976 he retired as captain of the US Navy, but stayed at NASA and helped with the Space Shuttle program. He retired in 1977 and continued to work as one of the leaders in the coal mining industry. He died in 1990 after a stroke.

Picture 50 Last picture from Apollo missions

2.6.10. Accidents on the Apollo missions

1. **Apollo 1:** The first test flight of the Apollo spacecraft was intended for a basic test on the journey around the Earth. The three astronauts, Grissom, White and Chaffee, boarded the command module. Concerned about the fact that they noticed full of loose intertwined wires, the management nevertheless approved the start of the preparations for the flight. Because the command module was hermetically sealed, they pumped so much oxygen that overpressure was created. Before the flight and while switching on the instruments, a spark was formed between the wires, which caused that oxygen immediately caught fire in the cabin. Despite the prompt call of the astronauts to save them, the rescue workers could not reach them in time. They spent several minutes to open the door of the command module; however it was too late for their rescue, and the astronauts died instantly.

2. **Apollo 8:** The first flight of Apollo with manned crew around the Moon. They did not have the lunar module with them that could be a rescue spacecraft in case of danger if the main engine of the command module would not work. The ignition and the power of the rocket engine had to be up to a second precise. If the thrust had been too strong or too premature, the rocket would have taken the astronauts past the Earth, and if the engine had not even ignited, they would have remained in the lunar orbit for eternity.

3. **Apollo 12:** Shortly after the launch of Apollo 12, a lightning appeared which was triggered by the Saturn 5 rocket itself. The lightning took the fuel cells for the

production of electricity to drive instruments in the command module offline. The second lightning directly took offline some key instruments. Spare batteries at the start were not so powerful as to enable all the necessary electricity for the instruments, so they stayed offline at the most crucial moment. Imagine, you are on a rocket, which swallows 15 tons of fuel per second, accelerates to 10000 km/h and you have no instruments. At that time, 24-year-old computer engineer John Aaron recalled doing a test in the event of loss of electricity in case computers remain without electric power supply. Astronauts had to find the AUX button to completely switch off all instruments, which is a very demanding task according to the size of the control panel. The proposal was sent to the astronauts; fortunately one of the astronauts remembered where this button was and then switched on the thing once again.

4. On April 11, 1970, **Apollo 13** was the third mission for landing with manned crew. Two days later, due to a mistake in electrical wiring, an explosion occurred, causing the service module to begin to lose oxygen and electricity, which prevented landing on the Moon, Additionally, it caused problems with navigation that was also dependent from electricity. The crew found shelter in the lunar module, where it remained until the return. This move was uncertain because the lunar module was designed only for the stay of two people for a period of 3 days, but to the surprise of all, this move turned out well. The command module remained operational, but shut down to save electricity and oxygen, thus retaining the ability to enter the atmosphere. Nevertheless, the crew successfully circled the Moon and finally returned to the Earth.

5. **Apollo 14**: Almost wouldn't have had a successful docking between the command and lunar module. Six attempts and nonstandard method were needed. They also had problems with the software and the landing radar.

6. **Apollo 15:** Also had problems with the Saturn 5 rocket engine, with an almost interruption of the thrust during the flight. On the lunar module on the Moon, the oxygen was leaking through valve. During the descent of the command module into the Earth's atmosphere, one of the parachutes did not open.

7. **At Apollo 16**, the takeoff was almost cancelled due to the damaged tank, and the guidance system wrongly decided that it could not navigate and demanded that navigation was transmitted to the reference points on the Moon and the Sun. The command module had problems with the control so that the descent to the lunar

surface almost did not happen and the crew would need a lunar module to return to Earth.

8. **Apollo 17:** Thirty seconds before takeoff, the computer interrupted the start of the flight, however, with the intervention they took off.

Apollo missions and their accidents were a great lesson in engineering. After the Apollo 1 accident, they completely changed and improved safety systems before taking off, and after the Apollo 13 accident, it turned out how the human mind can solve the most complex situations.

In the 1970s RCA had an ad that said: «The best electronic brain is still a human brain." Despite the media pomp about artificial intelligence, people are still our best problem solvers. Is it tempting to automatise everything? When it works, it is alright, but when it does not work, it is always a man's turn.

Conclusion:

All missions from Apollo 11 on advanced in their productivity; they learned how a man can actually operate on the Moon. Apollo 11 set the milestone and confirmation that a man can step onto the Moon, walk and perform simple tasks. The mission of the Apollo 12 had many more tasks, moonwalks were longer, they were exploring a wider area, but such a way of walking, climbing and carrying things proved to be a great exertion due to restrictions. The Apollo 14 crew already thought about a small cart for transporting tools and devices. It was the last Apollo mission with original technology. Apollo 15 represented a huge step forward, which brought lunar vehicle or (LVR). With this vehicle, they achieved greater distances, explored wider areas, especially remote craters, and did much more scientific tests without exertion. Apollo 16 and 17 represented only an upgrade and even greater productivity of experiments and researches than Apollo 15. It turned out that a man could work and use tool on the Moon. He also could live longer period if he had enough oxygen and food with him. This is already the task of future missions that can only upgrade this historic, pioneering epic journey.

We have passed through all the development process that described the greatest technological journey in history. What now? Is this explanation not enough for us? Is it possible with 350,000 people who worked on this project and billions spent that this journey could even be imaginary? Impossible. Nevertheless, for the greatest doubters, the explanation is given in Chapters 3, 4 and 5. Let's proceed and answer any doubts and explain all the claims.

3

Conspiracy Theorists on Moon Landing

3.1. Emergence of conspiracy theory

HOW HAVE WE come from one of the greatest achievements in the history of mankind, where almost one million people participated who did their best, to the conspiracy theory that landing on the Moon did not happen? On this topic, you can find a "false landing on the Moon" on Google with almost 700,000 hits.

Who created the conspiracy theory and why? Many people do not believe in anything, for example, a high school teacher who did not believe that the sinking of the Titanic in 1912 really occurred because she stated that nothing so big could swim.

Some people said before 1969: "You cannot just fly to the Moon, because the Moon is not just a light in the sky, it is a place." People, who cannot interpret the science of this, cannot understand. There are also people who would like to be heard and popular when they talk about subjects which are polar opposite of the actual situation.

We have two bases of theorists or conspiracy theories:

1. The conspiracy theorists during and after the development of the Apollo program

In late 1969, scientist (technical?) and writer John Noble Wilford put the theory of a real Moon landing into question. This gentleman was a great sceptic and, after an interview with Neil Armstrong, put the real landing into question. Mr. Armstrong, apparently, was not answering convincingly. This Sir was in a foreign world, he went through such stress like we cannot imagine on Earth. His condition would probably be understood only by psychologists, why he was answering as "earthling" unconvincingly. Or Mr. Nobel wanted to seem unconvincingly. Imagine that you cut off the weight of a $ 25 billion budget with one interview, in one newspaper. You must become a hero and a rich man for all eternity. It may have been the principle of this writer.

Picture 51 Statement of Noble Wilford

Picture 52 Skeptics at The Atlanta Constitution

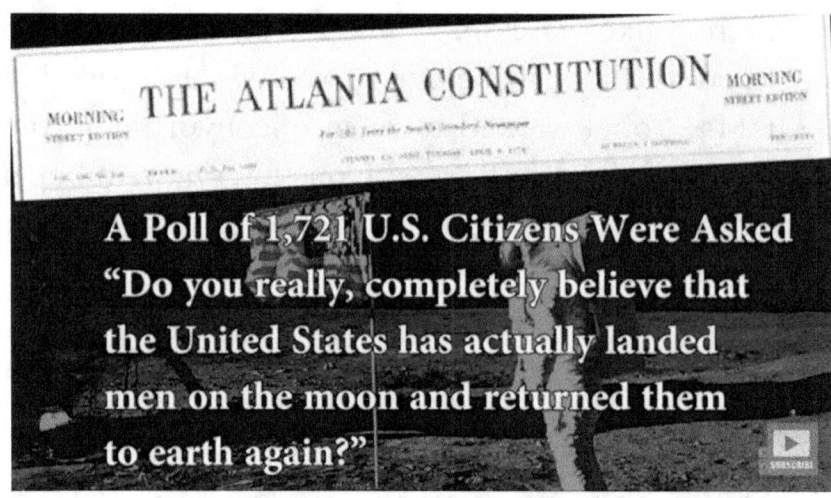

Picture 53 Skeptics at The Atlanta Constitution

Bill Kaysing

He was scientist at NASA, employed as a technical publicist for rocket engines. Already in 1963, he claimed that the F1 engines for Saturn 5 were not reliable enough for a flight to the Moon. He showed some documents to the public, indicating difficulties in the development of these engines. Such documents do not yet mean that the thing is not feasible. As an engineer, I can say that it can be hundreds of such development documents, and these do not prevent you from coming to the right solution later on. Nothing that has been proven to work was developed without such documents. Such claims are pointless and have no basis. These engines were developed at the end of the 1950s at the Rocketdyne factory, which were only upgraded for the needs of the Apollo program.

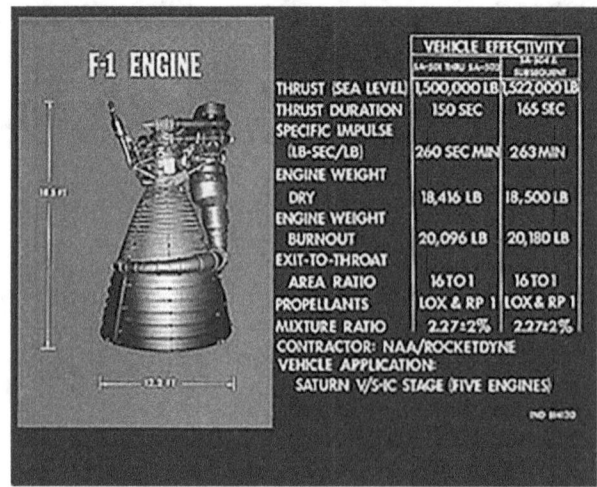

Picture 54 Rockedyne F1 Engine

He even argued that NASA did not use F1 engines but modified F1 engines to launch the rocket. Due to unreliability, smaller B1 engines of five, which were likely to built into the combustion chamber F1, may have been used. During the launch of Saturn 5, it is clearly visible that one flame is burning out of the chamber, and there is also a reliable level of rising and acceleration of the rocket. That this theory was fictional, the Atlantic Ocean was searched in recent time where the separated first stage of the rocket fell into the water, and from where the original F1 engines were actually pulled out. Keysing was not qualified enough as a professional because he was not an engineer in order to evaluate such technical conditions.

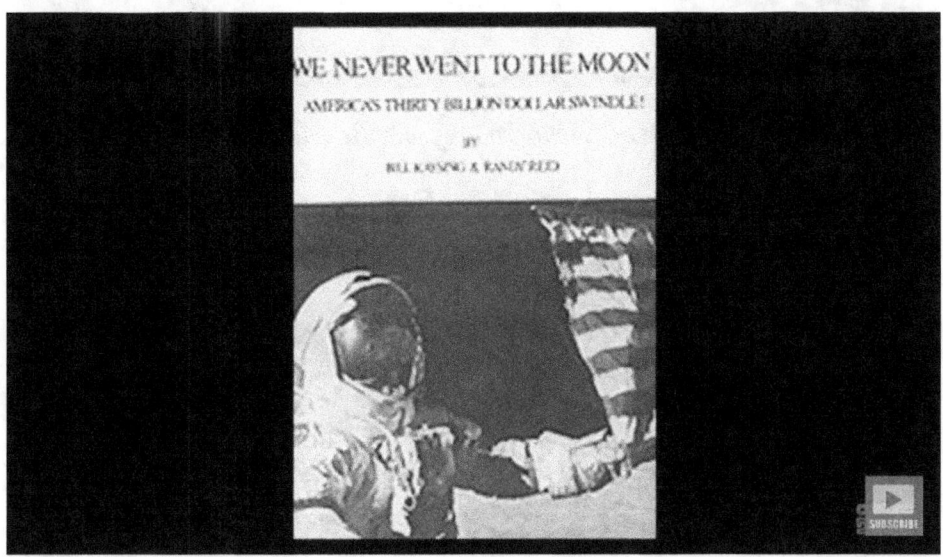

Picture 55 First book from Bill Kaysing

Problems with the F1 engine really existed, but they were later solved with a dividing plate on injectors and with additional injector holes to prevent a random pattern of fuel leakage from the injector holes. This random sample was caused by uncontrolled combustion and unequal pressure. The same problem was encountered in the B1 engines, on which they implemented the same solutions as on the F1 engines.

Keysing was responsible for reporting to the Congress on every development progress. His ideas of understanding the problems are absolutely nonsense from the engineering point of view. Keysing exploited the lay public, he was inclined to sensationalism. He has also changed his own story several times, which can also be proven. He was writing stories about the power of pyramids, the ancient astronauts. It was sad that people did not differentiate between reasonable conflicts and nonsense. Even today, it is not any better; at the most, worse, and it is even sadder that such people have the same influence as the "others".

Picture 56 Second book from Bill Kaysing

Stanley Kubrick

Picture 57 Stanley Kubrick and movie scene

Stanley Kubrick was an acknowledged film director who worked on SF films. His film Odisey 2001 was released in 1968. The effects of the film seemed real, but they were, of course, shot in the studio. He knew lighting techniques and special photographic techniques, contrast and focus editing. NASA hired him to illustrate the landing on the Moon with these techniques on Earth if landing on the Moon had failed, so that the world would not have watched in live if the astronauts had died on the Moon. Of course, Kubrick was shooting a film about Moon landing, for which he later said was not shot for the deception of the Moon landing. This story was also told by director and author Jay Weidner and he said that he believed NASA landed on the Moon with a human crew. The conspiracy theorists rearranged Stanley Kubrick´s statement, where he declares that he made images of landing on the Moon for NASA on Earth. In this publication, Stanley's family said that the interview on YouTube was a hoax, with the actor playing the film director Clockwork Orange, and that Stanley Kubrick was never interviewed by T. Patrick Murray. A two-hour film, supposedly a rough recording of an interview with Clockwork Orange director in March 1999, became useless only for a few days after NASA announced that it had found the crash site and parts of the Apollo 16 spacecraft on the Moon.

Who does not believe in such achievements? Those who are allowed to do so by their own ego have no visible success in their fields, their name is not known, their success is comparable to their abilities and they would like to be heard or to appear sensational. It is easiest for them to deny a theory with a mindless statement and questionable argumentation.

When perceiving this event, education has a major influence. Among those with higher education, 43% do not believe in landing, and only 23% among postgraduates.

54% of interviewed African Americans do not believe that humans landed on the Moon.

There are people who do not believe in anything and anyone; such people reject everything or mostly that is technically provable. Here we meet with the fire effect. The more you fear the fire and want to extinguish it, the stronger the fire is. So the more we want to prove the sceptics they are wrong, less they believe.

In 1964, 76% of Americans believed the government that it was doing what was right; in 1990 they were only 25%. The reason for this is the various affairs that the government has had. It began with the Watergate affair, terrorism, with the peak of the Cold War, sending Americans to foreign battlegrounds. All of this has affected the fact that fewer people have trusted the government and everything that was connected with the government's activities. Therefore, in recent times fewer people believe in Moon landing than before. This should not be related to scientists, engineers and all the other administration that allowed landing on the Moon. These are not politicians, they are people with dignity. Everything they do, they do it carefully and convincingly.

It is psychologically proven that people with poorer education, whether this be Moon landing, general politics and various predictions, etc. do not believe in big stories. Landing on the Moon was a very demanding project and was also enabled with help of the ingenious human mind. Why do not these people believe it? It is almost impossible to change their mind because they have no basis to understand the scientific explanation of the opposite. More than 63% of ordinary people who vote for general election to the Congress have responded that they believe in at least one of the conspiracy theories, and if they believe in one theory, then will also believe in others.

Today, times are much more unstable than they used to be, so this affects people, poverty, unpredictable future, and finances. In this state, people are much more susceptible to conspiracy theories. The Internet, with many unverified data of contradicting facts, only deepens this psychological crisis of mankind.

In the future, no matter what NASA or other space agencies will do, there will always be people who will not believe anything. They themselves do not want it or they are not able to check it.

2. New conspiracy theorists have emerged in today's generation.

Why is there a doubt? The world has evolved from euphoria and success into a crisis. In 1973, oil crisis arose, which for the first time paralyzed the normal life of prosperity, self-evidence and accessibility to everything. With the arrival of new strategies for the great powers, the tension has increased. Towards the end of the 1980s there was a new prosperity, the countries' GDP was growing, people wanted to make a new life, and live better, and forget about everything what was bad. Space agency budgets have declined, and NASA, however, successfully launched Space Shuttle in the early eighties. These were wild years, new inventions have come, computers, entertainment, tourism were on the rise and socialist governments were falling in Europe. In a quest for a better life, Europe and the whole world was captured by even greater globalization. At the beginning and the end of the nineties, the world was worried about the transition to the new millennium for the number 2000, how this number would affect computer systems and the Internet. Many systems of national importance have been computer-controlled, and in case of a system crash, large-scale disasters could have occurred. The world was tense, local wars were raging, as well as problems in the Middle East, and there were riots in various countries. Everything has somehow stopped and a lot of hope that it would be better for everyone may be enabled by digitalisation in all spheres of life, which would also result in reducing of working time. The old and new superpowers were seeking their role in this chaos. Expenditure on military budgets has increased greatly, at the same time the gap between the poor and the rich widened. Someone saw the opportunity to arrange the world differently to have influence in all spheres of social life. These were bank corporations, multinational corporations controlling the production of seeds, and other guards and watchmen of money, and tax haven played into their hands with that. To achieve these goals, large investments in digitalisation go hand in hand. Due to the need for greater digitalisation and personnel that would enable this, the millennial generation was created. This generation is convinced that it is the milestone of all new things, while forgetting the exceptional and bold achievements in the past. They believe that they are the ones who are creating a technological breakthrough in history, although computers were invented in the sixties and Wi-Fi in the forties. Regardless of the fact that it is possible to store many more useful applications in a smaller processor space today, the processors' tactic has been the same before and now. New conspiracy theorists began to appear, who do not believe everything that happened before them because of the misunderstanding, the gap and the disconnection among achievements in the 1970s and the latest technology in all spheres of social life have been achieved. The myth has been created that without digitalisation nothing can longer be done, and that everything that has happened before is not to believe it, would work successfully. Even a part of the highly technical personnel from the computer field began to doubt, which themselves should be able to reasonably explain some questions and explain the doubts that arose. This superficial thinking caused a lot of damage.

3.2 Opinions of ordinary people who do not believe in landing on the

Moon

1. Claim

MAN NEVER LANDED ON THE MOON!!UNDERSTAND THIS ONCE AND FOR ALL...everything is stage-managed and an intrigue that NASA has been trying to hide until today...why no one has returned back, we will never know...if we cannot come to the Moon with today's technology, then I think that all was a hoax for the naive nation...if today's mobile phone has greater capacity of operations than all the computers of that time, then I think we all have the answer in front of us... The Russians did not go to the Moon why? Because they knew that radiation would have killed them after the first 20000 km of space flight... And all of you that believe in landing on the Moon are nothing but an ordinary sheep.

1. Explanation

Not that in today's theory we cannot get to the Moon. Of course we can. Nowadays, there is a complete non-visionary atmosphere in the world that does not know what to do with itself. The advanced computer skills themselves are not enough. It is about special mechanical, gravity, physical questions. The mathematical explanation of these skills is more than possible and has been available for centuries. These are values and principles that are not on the right track now.

It is not true that today›s mobile phone has more capacity than all the computers of that time together (see Chapter 6). Today›s mobile phone is not nearly as capable of 3D design as the graphic workstation from the early 80's. Today›s mobile phones were given more options for more applications. You do not need these fun-filled applications for a flight to the Moon.

Radiation does not kill you; it is manageable, as it is not mentioned as a problem during the preparations for a human flight to Mars. The Russian male and female cosmonauts from the middle of the 60s, who circulated the Earth at least 20 x in the Vostok spaceship, later had children and there were no consequences because of radiation. The child was completely normal.

2. Claim

In 1969 there was a great technology for film making. In that year, two classic "Hello Dolly" and "Butch Cassidy and the Sundance Kid" films were shot. The quality of these shots is not comparable to the quality of the images on the Moon. Do you go to the Moon with a camera of the worst quality, which you can buy at the news-stand? No logic!

2. Explanation

Explanation: See the quality of the film "The Wizard of Oz" from 1939; it will be all clear what kind of quality was possible already at that time.

3. Claim

"Why, today, 43 years after the first landing, we do not go to the Moon if the average laptop has a higher capacity than the Saturn 5 rocket, and the USB key has more memory than a computer on the Apollo capsule which was floating to the Moon?"

3. Explanation

It's not true. This is not about the memory size at all. Another calculator can calculate the trajectory according to the most demanding mathematical formula.

There is a motherboard that led the launch and the rocket path, which is comparable to today's capacity and size. A comparison that one laptop has more capacity than the computer at that time does not say anything. The point of space flights is rockets and their ability. Calculations of the necessary paths for motion among the planets have been known for almost 200 years. Moreover, when and for how long which stage of the rocket switches on, does not require a large computer.

4. Claim

When General Vernon Walters (Director of the CIA) was interviewed in 2001 for the purpose of a French documentary, this was at the same time his last interview, because he died for some reason the day after. It was fatal for him, since the evening before he had already indicated his intention to tell the names and plans behind the camera, who did the things and how. He was visibly confused and warned the French ARTE team to turn off the camera and that it could cost someone›s life. He indicated that he would break one of the CIA›s explicit rules of "secrecy and anonymity".

4. Explanation

Why did not they kill all 10 astronauts who landed on the Moon, since they are the biggest witness for or against?

5. Claim

Nixon and his magnificent five advisors (Donald Rumsfeld, General Alexander Haig, Richard Helms (Director of the CIA), Lawrence Eagleburger and Henry Kissinger) became scared and began to think that protagonists should "disappear forever" and this should mean exactly this. Nixon sent 150,000 soldiers to search for four fugitives. The sound engineer was burnt in a car accident. The assistant director drowned in his own pool at home in front of the house. Another engineer was found in pieces in Patagonia. The CIA recruits even recorded his murder, somewhere in the Kerguelen Islands. The set technician found out what was going on and he was hiding for 10 years in a Jewish monastery in Brooklyn until they came to him. Thus Stanley Kubrick remained as the last surviving, last living witness to the project of studio recordings of landing on the Moon. He decided to disappear. He locked himself on his estate, along with his family, and never went out again until 1999, when in the strange circumstances he died of a heart attack immediately after the release of the controversial film Eyes wide shut.

5. Explanation

This may be unverified fiction. But did someone miss a friend who would disappear in a suspicious way? Nobody missed anyone. It would be rather suspicious if only those who worked on the studio scene would be killed. Where are the other 300,000 people, scientists and engineers who believed what they were doing, because they could regularly check the outcomes of the development and see what was happening on their own.

6. Claim

Henry Kissinger and Donald Rumsfeld, the then adviser to Nixon, visited the MGM studio in England the same evening where they just were finishing with filming Space Odyssey. "Stanley, we ask you just one thing. Just leave us the keys from the studio for a weekend to take a few scenes and to take a couple of pictures; everything will be cleaned by Monday morning." A film with four skeletons of CIA agents, two astronauts and two technicians, was intended for filming. All four were single, without a family, and all signed an agreement to not disclose information and the condition that they "disappear forever" after the recording was finished. At all, Stanley Kubrick was present as the main consultant for making the recording.

6. Explanation

Kubrick later said that he was doing scenes for the film, not for a real landing on the Moon. Over 16 billion dollars and 10 years of hard work of hundreds of thousands of people should not go to waste and therefore, in any case, if the landing had failed and if they had died on the Moon. Or if they had lost signal and picture, they could have replaced this outage with studio production, even though everything would have been normal on the Moon. Only for this they were asked for a studio scene, which was prepared.

7. Claim

Vernon Walters (CIA Director) said in an interview: "I warned Nixon that it was very dangerous to lie in the United States. And to carry out this kind of deceit in a democratic society is too risky. Too many people would talk, it would become absurd. "Nixon replied with a sad voice:" Anyway, go on!"

7. Explanation

Venon Walters had in mind if, in the event of an accident or a loss of a signal, he would carry out a studio scene and would actually portray untrue recordings of how the landing took place in the finest order but not the actual ones if an accident could have occurred. If an accident occurred in a live transmission, it would still be a shock. Imagine if your friends or relatives watched their dear ones dying.

8. Claim

After the finished mission, the first man on the Moon, Neil Armstrong (born August 5, 1930), moved away from the public and went to the monastery. Michael Collins, the third astronaut, as the only one who did not walked on the Moon, also disappeared from the public. Another man on the Moon, Buzz Aldrin became an alcoholic and fell into deep depression. "There have been some unusual things that surprised me and have had a great impact on my life before we flew to the Moon," says Aldrin. He knew that President Nixon made a secret speech a few days before the flight, announcing the death of the three astronauts if they had not returned successfully from the Moon. Aldrin concludes: "When you are suddenly faced with such an adjustment of the situation, you change, or you die. That's why I decided to retire from NASA and the army... Did we, did a man go to the Moon or not? "... he also said at the end. A strange statement, a questioning phrase that originally read: *"Did we,...did people go to the Moon or not?"*

8. Explanation

The first claim that Neil Armstrong went to the monastery is a lie, so others are not worth commenting. Neil Armstrong went to teach aeronautics to the university. In the 1970s, he even went, with the researchers of National Geographic, to explore a very deep cave in South America, where copper artifacts, which were left in the history by foreign civilizations, supposed to be located. I saw the film with my own eyes, as I saw Neil Armstrong. It is true that the journey to the Moon influenced him; these are the experiences that we "earthlings» cannot imagine. He needed some peace from the public, but he did not retreat.

9. Claim

Interesting, isn't it? Now, with all this technology, we're not getting any closer.

9. Explanation

The superb technology, the system how to get a man to the Moon (stages and thrust of rockets, path trajectories) took place in 1969, of course, with the support of the smaller computers than today, but they were good enough and reliable. This technology (except computers) IS ALWAYS SUPERB, and it will take 100 years before someone finds better drives for direct landing and takeoff on other celestial bodies.

10. Claim

And, how the flag fluttered there in such a rare atmosphere without wind, didn't it!!!;-) Prank, duuuuh!!!

10. Explanation

It is not worth answering such an ignorance of the doubter.

3.3. Thoughts of people who believe in landing on the Moon

1

It is sad...the fact that folk is becoming more and more stupid (evidence is already a massive participation in the elections, if not the other... As for the landing on the Moon, there is of course no doubt that it happened...there is also no doubt that they had a backup version recorded in the studio (in case the landing would not be successful), in which case the recording would have been included in the direct transmission, and the accident would have occurred on the way back. At that time, America simply could not allow failure—therefore, a backup scenario that now disturbs many "hobby conspiracy theorists".

What is essential, however, is not known to everyone, or not enough attention is paid to this...

The Apollo project was terminated prematurely, despite the fact that spacecrafts were already built, finances were provided, etc.

The fact is also that the photos from the Moon have been retouched (just why :)? And the biggest fact is that nobody ever dares to land on our beautiful satellite, despite the enormous advancement of technology in these years. And besides, we are attacking the Earth>s orbit for an eternity...

Those who know little physical laws is like driving from New York to Los Angeles, and then turning back before San Francisco because it would be too expensive to LA...

2

Well, come on! Why are you so skeptical? If you have your problems, do not doubt the things that are real. I'm sure that on July 21, 1969, a man landed on the Moon!

3

I'm sure they have truly 6x successfully landed on the Moon. More than 100,000 people participated in the lunar program, and surely we would find someone among these who would be willing to tell something to the eastern block at the peak of the Cold War for megabucks.

4

I'll just say that. I will not claim that it is true and I will not claim that it is a hoax, but I still think that if it was a hoax, the whole world would know it because the Russians would compete with the Americans and they had spies all over the world, and this would 100% come to their ears etc...

5

Perhaps many do not understand what did the conquest of the space meant immediately after the end of the Second World War, when the radar, encrypted radio signals, rocket engines, new composites, various robotic mechanical devices, life-support technology used in submarines, a new type of canned food and suchlike were used for the first time. Before the Russians launched the Sputnik into orbit, they had assumed that the space is a world border, like the legend of the end of the sea, where water flows into the void. Only then did the awareness of a round earth began to emerge which is not hanged on anything, in the middle of the space void, whatever it supposed to mean. In those few years of war, more science has developed than before in the centuries, a global scientific community has begun to form, which redefined the millennia of accumulated human knowledge of various cultures and civilizations. Today, just half a century later, we take a lot of things for granted. We see the first space flight as prehistoric, although most of today's technology is actually an upgrade and development of already existing technology, computing and information technology, biotechnology, nanotechnology, psychology. All of this has already been in the 50s, even complex devices such as microprocessors, robots, materials with memory; fission and fusion, genetic modifications, most of the innovations around us are the result of the evolution of the science of that time. The question of what innovations are exclusively belonging to today, what has changed in the past 50 years? Today, we live in idiocracy, although almost no one understands anything, because of the development of humanities and culture, we nevertheless seem to be extremely smart and intelligent, but it is not true. Today's experts are much better organized and equipped, knowledge is highly efficient systematized, but in comparison to those engineers who were also top-level and highly disciplined philosophers, historians and artists, we are actually monkeys. It seems to us that we are extremely effective, but this is far from being true; it's a greater miracle to travel into space today. I wonder what else they discovered at that time, which still cannot be discussed by the human community, because the very foundations of morality, ethics and society could be shaken, and then we would only understand what are these real conspiracy theories about God, the space, life, and everything in general. If I wrote that life is "xnijhjnguk", it's probably clear to everyone what I meant by that, isn't it? It isn't? Some things are simply unspeakable; there don't even exist words and concepts yet that would imply a non-empirical idea, which in fact happens. First, we have to invent them; only then the science will be able to make progress again.

6

99.98% of the commentators are illiterate. I claim that in 1969 Neil Armstrong made the first step on the Moon. Although I do not support the US in a military sense, but in conquering the space, I have to say that they are excellent; also the Russians are cosmic old hand and everyone else then developed technology more easily. So there is no dilemma, man walked on the Moon.

7

I pity everybody who does not believe in landing on the Moon, really. During the space race, when all the satellites of Russia were targeted at Apollo, were they supposed to come up with this? And the Russians saw nothing and admitted defeat? Comedy of the millennium... So the space station MIR is also a fiction, as well as a Hubble telescope, and GPS is basically radio stations on the hills?

4

The claims of conspiracy theorists and explanations

4.1 Fluttering of the flag on the Moon

IT IS EASY to explain that the flags were fluttering on the Moon. If an object in the space or in vacuum obtains acceleration to a certain speed, the body will keep this speed infinitely, until it is prevented by another opposite force. When planting them on the lunar surface, the flags received a certain amount of momentum. This momentum had an influence on the movement of the flag, which, however, lasted much longer than on Earth, because the flags, unlike on the Earth, had no air resistance. The recording of the flag fluttering when taking off of the lunar module from the lunar surface was caused by the exhaust gases of the rocket engines. The picture below shows the two edited recordings in a different period of time. This picture serves as evidence that the flag did not move in two time periods.

Picture 58 The flag before takeoff of the lunar module

Picture 59 Fluttering the flag after takeoff of the lunar module

Picture 60 Comparison of the pictures in a minute interval

4.2. Non-parallel shadows

At such a great light source as the Sun, shadows of objects on the surface should be parallel. However, since the surfaces themselves on the lunar surface are not completely straight or parallel, it means that also shadows cannot be parallel despite the great light source such as the Sun. In the picture below, we see the simulation of non-parallel shadows of the same light source on Earth.

Picture 61 Four pictures on top, the combination of the rough surface of the Moon and the low Sun can easily create a complex shadows.

Picture 62 Shadows on the Moon

4.3. The stars in the picture cannot be seen

The most common question of conspiracy theorists is why stars in the Moon's sky cannot be seen. The answer is quite simple, and with a little logic every average person can get the answer. Even on Earth, when we take normal pictures at night, we cannot see the stars in the pictures. It is a matter of time exposure of the shinning object we are taking picture of. We have a combination of very bright and less bright-colored objects.

If you set the exposure, in order to see less bright objects, the brighter objects would be very shinning and would not be able to distinguish the picture. Therefore, the photos were adjusted to the light of the lunar surface, which is much brighter than the stars in the sky. You also cannot see the stars in the pictures of the Earth and the black sky from the international space station.

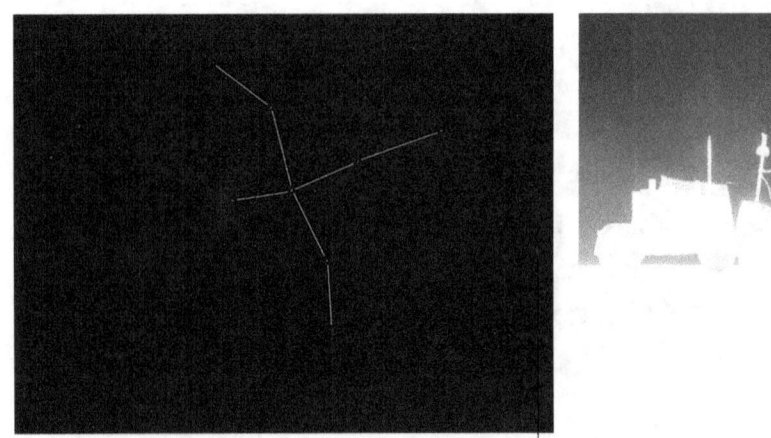

Picture 63 Comparison of the same dispositions of the sky with the stars and a lunar vehicle

Images show photos with the same time exposure. In the night sky, we can see pale stars, and the lunar module (used as a model) was illuminated with the same light as on the Moon. We see that the lunar module would be too shinning during this time exposure.

Picture 64 Comparison of the sky on the Moon and the sky visible from the international space station

4.4. Van Allen belts

In the space there is the Sun's radiation, from which we are not protected, unlike on Earth. There are two Van Allen belts that surround the Earth, a large and a small circle. These are the two magnetic fields created by the Earth's magnetic field. These fields are enhanced by high-energy particles from the Sun, so-called solar wind, which causes radiation.

The theory of opponents says that people cannot travel through these belts because they would be killed instantly by radiation.

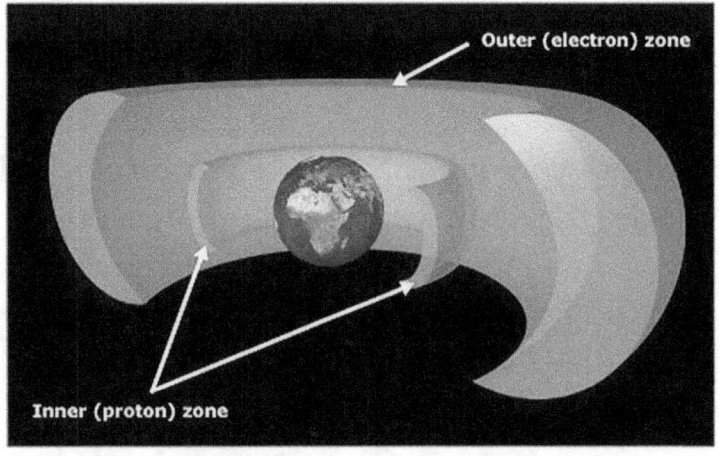

Picture 65 Van Allen belts around the Earth

The problem is how people understand the concept of radiation. Most of them immediately think that radiation is a mystical force and immediately causes death. Radiation is the decay of the atomic nucleus. In this process some energy is released in the form of kinetic energy of protons and neutrons. The astronauts' journey was planned in such way that they passed through the energy weakest parts of radiation. Nevertheless, the astronauts had some protection. The first one was a protective metal plate on the spacecraft, and the other was the astronauts' spacesuit itself consisting of 27 parts of protective materials. The consequences of radiation are also affected by the exposure time. Astronauts spent 4 hours on crossing the belts, and they were traveling for an hour through the most dangerous part. The scientist who discovered these belts said that it would kill a man if he was exposed to them for a month.

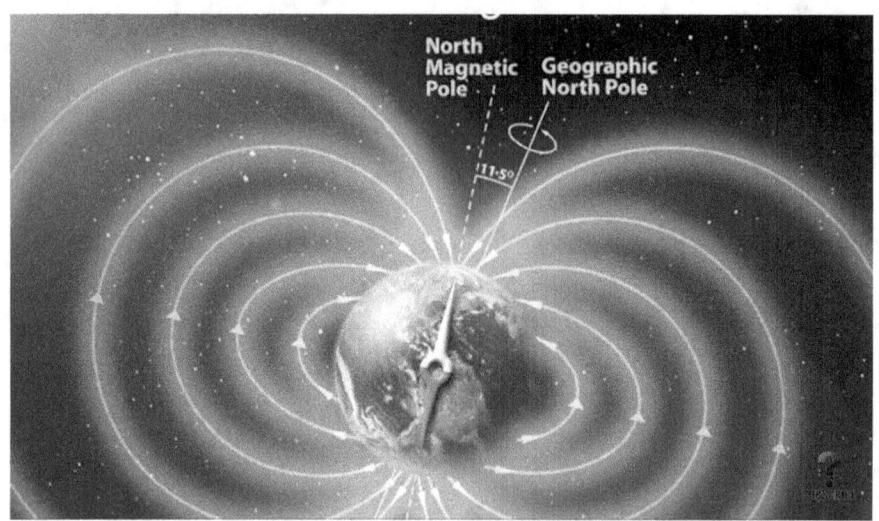

Before the Apollo mission flights, NASA had been aware of these energetic radiations as they were discovered by Explorer, Pioneer and Moon in 1950s. The intensity of these radiations depends on the Sun's activity or the solar wind. Fortunately, Apollo's flights were planned when the intensity of the Sun was at a very low level.

Radiation related illnesses would have occurred if they had been exposed to 200 to 1000 rads. During the Apollo journey through these belts the intensity of radiation was only 18 rads, that is, within the safe area. However, the Apollo spaceships were additionally isolated, so that the astronauts received a dose of radiation comparable to the chest radiography. The average dose was even 0.18 rads due to these measures.

Average Radiation Doses of the Flight
Crews for the Apollo Missions

Apollo Mission	Skin Dose, rads
7	0.16
8	.16
9	.20
10	.48
11	.18
12	.58
13	.24
14	1.14
15	.30
16	.51
17	.55

Table 5 The average radiation doses during the Apollo flight

4.5. Mismatch of footprints on the Moon with the soles of the spacemen

Picture 67 Comparison of the astronaut´s suit with the footprint on the Moon

This is another fictional, very simple story. In the footsteps, it is seen that this is just a spacesuit, on which the spacemen also put space shoes as shown in the picture.

4.6. Light on the astronaut´s spacesuit in the shadow

Many people were convinced that the illuminated spacesuit in the shadow was due to the extra light in the studio. The lunar surface is so shinning that it radiates light on the bright space suit, which can be seen as a reflection of light. It is the same example as on Earth when the full Moon illuminates the landscape at night.

Picture 68 Reflection of light on spacesuit

4.7. Footprints on the Moon surface

If we make a footprint on the sandy ground, it is seen only if sand is sufficiently humid. On the Moon, astronauts made visible footprints. For this reason some people thought that the footprint was falsified in the studio. Due to the erosion the Earth's sand had been shaped into a smooth shape of particles for centuries. The pressure on the smooth particles does not keep the shape that was made because there are no sharp edges that would hold the particles together. There is no wind and water on the Moon, so the particles (regolith) have remained for centuries with sharp edges. Due to the sharp edges of these particles, the footprint caused the fine sand to retain the foot- printed shape after the pressure.

Picture 69 Footprints in Regolith

Picture 70 Comparison of normal sand from the Earth and Regolith from the Moon

4.8. Why the exhaust crater cannot be seen, and why the feet are clean?

The conspirators used evidence that the photographs are not from the Moon on the grounds that the exhaust crater cannot be seen under the lunar module. Theorists have shown that they do not know physics. The engine of the module has the maximum thrust of 4.5 tons, and the used thrust was up to six times smaller because on the Moon there is six times less gravity than on the Earth. There is no atmosphere on the Moon, and that is why gases are spreading much faster than air. Vacuum dust also falls on the ground faster than on Earth and does not swirl, so it behaves like small rocks on the Earth. Therefore, we see only a small rise of dust under the exhaust nozzle in the picture.

The conspiracy theorists are interested in why feet of the lunar module are clean and not covered with dust. Before the landing the lunar module has probes lowered, which first touch the ground and turn off the engine. Just above the surface, the engines are turned off. During falling with the engine stopped, dust is already lying on the ground, before the feet touch the ground.

At the landing site of the lunar module of Apollo 11 it was a very thin layer of dust on a relatively solid surface.

Picture 71 Display of the exhaust crater and landing leg of the lunar module

4.9. Rise of the Earth on the Moon

The conspiracy theorists displayed an animation that shows the Moon how it always faces the Earth with one side while circling around the Earth. It is true that the Earth and the Moon have a special gravitational relationship. This means that the astronauts on the Moon would not be able to see the rise of the Earth over the Moon's horizon, as shown in the films. Rise of the Earth was never shown while the astronauts were walking on the Moon, but during the circulation of the command and service module. In this case, the Moon's horizon is slightly bent, so the film was shot from high altitude. Given the speed of CSM circulation around the Moon, it is perfectly clear that the rise of the Earth over the Moon's horizon is visible.

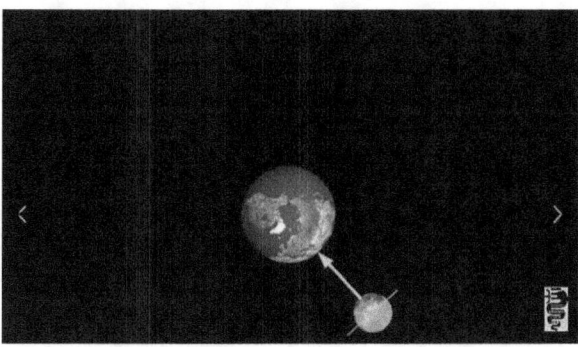

Picture 72 Gravitational relationship of Earth and Moon Picture 73 "Rise" of the Earth over the Moon

4.10. Photographs on the Moon

1. The conspiracy theorists say that the pictures below were taken with the scenery in the studio, because the "scenes" in the two photographs are the same, and the front frames are different. The scenery is not the scenery, and is not close but it is very far away because there is no air on the Moon and the impression of the distance of the object is wrong. If the hills are at least 20 km away, and if we take pictures of various things 100 meters apart, the background will remain visually unchanged.

Picture 74 Photograph taken from two different positions

2. In the bottom left figure, the astronaut took a photograph. The conspiracy theorists said the studio light was behind it. This phenomenon in photography is known as "Aurora" when the light from the photographed area is reflected back to the camera. Regolith, lunar dust has great ability to reflex light. The same example can be found on Earth, for example, in the morning dew.

Picture 75 Display of "Aurora" on the Moon and on the Earth

3. The conspiracy theorists claim that there is a reflection of the studio light on the visor. Every photographer knows that in the photo, one reflected light from a photographed object can be seen if there is a sun or some other reflection of the light source in it. On the right picture there is an example from the Earth where the photographer took a photo of an aluminum ball that reflects the Sun's light.

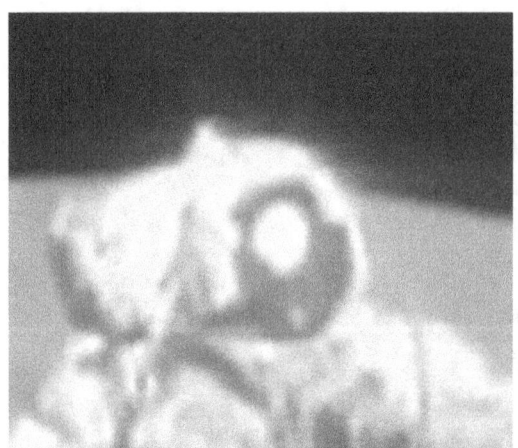

Picture 76 A reflection of light on a photo on the Moon and the Earth

5

Perception and cognition

S O WHAT DOES it take to believe or not to believe something? In this emotional response we rely on ourselves, with the data and the knowledge that is available to us. This response is influenced by past, perhaps negative, experiences, which give us a pattern of perception and thinking for all further opinions. The pattern that is produced is not always good enough to get to the truth. If we want to know the truth or try to find a way to get to it, we can turn for help in the philosophical field, to the field of cognitive theory or epistemology.

Sceptics have a question that comes back time and again: "Were people walking on the Moon in 1969 or not?"

For the ability and the explanation for you to believe that they were on the Moon, it still takes a lot of technical knowledge, and great confidence in the ability of scientists. If you do not believe in a reasonable explanation, you may not be able to use the right senses. Before perceiving a more detailed explanation, you must be absolutely impartial and not to cultivate any intimate resentment to either side or to be a follower of sensationalism.

When should we believe what others say? Do they often lie on TV or on the radio? We want to find out what information is real. Philosophy endeavours to reach the recognition that we would recognize what the world really is. Philosophers believe that before we begin to check the credibility of information on television or elsewhere, we must check whether we can trust our basic sources of information—that is to the eyes, ears and reason. Why would you doubt your own senses? Because our ideas often deceive us. Dreams sometimes seem like reality, or now when you are reading this are you really reading it, or is it just a fiction? Scepticism and doubt are healthy to some extent, but if you do not believe in anything, despite the evidence, it is nonsense. Also for reasonable evidence you say that they are adapted, so everything can be adapted and we do not live in the real world. Then we should not even believe despite family happiness that we have a family.

To some extent, there is a healthy methodical doubt that says almost all our beliefs are questionable.

So how can we assure ourselves? We can assure ourselves in two ways:

1. EXPERIENCE

2. REASON, MIND AND INTUITION

The question arises on which source of belief can we rely more and which is more important?

We have not any experience whether we are sure that they were on the Moon or not, but we have reason and intuition. Perhaps in recent times there has been a doubt because fewer people trust the governments than they used to, and that is why more people question the government's certain moves and link also the past with that, even though the level of confidence in the government was very high during flights to the Moon. So in this case, this experience does not bring any logical connection, so we need to rely on the reason. What does the shared experience of all the people who watched the launch of the rocket and landing on the Moon then mean? It is not necessary that the two people saw equally, although they are very similar in character, but differ in the fact that one sees a real landing, while another sees landing as a studio processing. Here too, we do not get to the logical conclusion. Everyone wants to explain why he thinks that way. There must be discussion and argumentation between two different opinions. I think that the pros are more rational convincing, which I will show in the next paragraphs.

The following cognitive methods can be used:

1. Induction

This is every conclusion that is not based on logical derivation, but on the study of individual cases. Before the flight to the Moon, many gradual steps were carried out; the flight was not performed all of a sudden. They carried out a lot of experiments, huge security measures, and tested every next step in detail. So we can conclude that it is unbelievable that they would not have made the last step, landing on the Moon.

2. Abduction

It means reasoning the best clarification. If more than 300,000 people have decided to carry out the project to the end and have done countless tests, it is not logical to overcome the obstacles that were solved during the development and then to depart from the final goal. So, we can rely with great certainty on the best explanation possible that the project was carried out to the end.

3. Deduction

Although this is a logical deduction from previous evidence, this method is rather mathematical and it is not fully an option here.

In all of this, relativism can lead us the most to doubt; therefore everyone can be right according to the way of their experience. It may be that both are right, or none, or only one. In this case we have to agree with each other. This again does not lead to the actual truth. It is even more illogical that someone who agrees with relativism, that all is possible, has to agree then also with the opposite opinion.

In this chapter I describe what reason and intuition mean. All other empirical beliefs that are most common among conspiracy theorists are merely a presumption. Let one of the conspiracy theorists prove me the contrary, with reasonable argumentation supported by physical evidence. However, he must be clean on his proving, and his "proving" must not be filled with certain interests or needs. This is precisely the case most often among conspiracy theorists. They like sensationalism, but it has very much contaminated people without proper prior knowledge and eager for sensationalism. People who did not have their opinion succumbed to the influence of sensationalists. The need for sensationalism and reason is inversely proportional. I have mentioned the details of this in the chapter Conspiracy theorists, where I was speaking about how many people believe in landing on the moon and how many do not believe, also in relation to education.

In the chapter on evidence, I tried to use the reason, because only this leads to real cognition. I chose the reason because the Apollo project is ideal for proving with reason, since the whole system is built from the facts and knowledge of thousands of scientists and engineers which can be explained easily.

Let this book be a good source of cognition and show what all those who doubt do not want to think about. Great things are planned step by step, and this book should present what can come from proper planning and vision. Let›s learn what the real sources of cognition are, how to reach the truth, with what kind of method, and what the basics for proper interpretation are. Let us examine Chapter Six.

6

Evidence of landing on the Moon

I AM GOING TO present the evidence based on reasoning with use of reason. Sometimes you have to go deeper and deal with facts from different point of view. Not everyone can be able to see the facts, which are presented evidence, as actual evidence.

6.1. The Saturn 5 rocket was so powerful that it was also planned for flight to Mars.

Werner Von Braun was the chief rocket engineer in the Hitler's rocket program where he developed V2 rockets that they used for the attack on London. As a man with the greatest knowledge of rocket technology of that time, he went to the United States after the Second World War. At the time, the United States had not yet thought about such, almost fantastical, journeys, such as the journey to the Moon with a manned crew. Werner von Braun had a lot of knowledge and experience, and the Americans knew in some way that they would benefit from this knowledge. When the space race between the Soviets and the Americans began, the latter then decided that they wanted to land on the Moon with manned crew before the end of the seventh decade. Werner Von Braun was a visionary, and he even had plans for the journey to Mars. The Saturn 5 rocket had been actually dimensioned for the journey to Mars, but was later developed for the journey to the Moon. The first forerunner of the Saturn 5 rocket was therefore a wartime V2 rocket. During the war, a V3 rocket was also developed that could carry a combat load even over the USA. Neither the takeoff of the rocket Saturn 5 nor its size was a fiction. The rocket and its takeoff were seen by everyone and they believed the rocket was there and that it was launched. The takeoff of the rocket and its orbiter into the Earth's orbit are the hardest part of the entire flight to the Moon from the scientific and engineering field. Everything else, deorbiting from the Earth's orbit, flight towards the Moon, the orbiter into the lunar orbit and landing on the Moon, is technically less difficult, but it represented a more difficult part in terms of safety. Each element or part was built in such a way that it served a precisely defined purpose. In the pictures below, we see a flight of the rocket at its full power. If a 110 m long rocket from the picture shows that the exhaust gases were more than 300 meters long, there is no need to demonstrate better evidence of the actual power of the rocket.

Picture 77 Flight of Saturn 5 at full power

The CMS (Command and Service Module) had only one thrust nozzle, therefore it was a need for the lowest-power drive. Each previous stage was stronger, more complicated, in which a mistake and an accident of a larger scale could occur earlier. The most powerful stages have worked and were visible from the Earth. Why would somebody succeed in the most difficult part, but not in the easier one? Why someone would not do an easier job if he had a goal just before him and spent billions of dollars? I see a less difficult part in terms of the need for smaller forces, as required when taking off from the Earth. Greater force means greater strength and safety challenges. Why would the Saturn 5 rocket be made of such complicated and heavy parts so that they could then broadcast studio recordings on television? It is like if someone drove from New York to Los Angeles and turned right just before Los Angeles and drove back because the journey would be too expensive.

6.2. Movement of people on the Moon and driving with Lunar Rover

A revealing recording is driving a lunar module across the lunar surface. Rover jumps and falls during driving on the uneven terrain with one sixth of Earth's gravity. When it drives onto the bump, it is lifted to the air as fast as on Earth, but when the vehicle is falling to the ground it is obvious that one sixth of gravity has an effect on it. At the same time, fine dust from the rolling tires can be lifted very high due to one sixth of the gravity. The dust falls relatively quickly because we also have to consider a vacuum in which there is no air resistance, therefore it is falling at the same speed as a large rock. Depending on the speed of Rover's driving on the Earth, the dust could not have lifted so high from under the wheel. So the drive took place on the surface where there is one sixth of Earth's gravity. The clarification of conspiracy theorists with falsified slow recordings (fast-slow-fast) is technically unfeasible and could not give such recordings as we saw them.

Picture 78 Photograph of the dust lift-off while driving the rover

On Earth, it would be impossible to falsify a recording with different movements, especially when these movements occur at the same time.

The second recording shows the astronaut's fall. During the fall, we notice that one sixth of gravity has an effect on it. When the same astronaut subconsciously reacts with his feet to defend himself from falling, he seeks balance with the speed of the movement of his legs as if he was reacting if he was on Earth. So when he moves his feet to save himself from falling, his legs do not move in slow motion. Again, we should falsify the recording (fast-slow), which is not possible, but if it was, the movie skips could be detected while changing the streaming speed. Two-speed movement cannot be adjusted with one recording. According to reasonable explanation, such movements had to occur only on the Moon. The astronaut's jump has also the same conditions. The push-off is seen in the normal, earthly movement, and the fall of the astronaut towards the ground is slower than on Earth. Even such a recording in the same time cannot be edited or divided. An exact view on the combination of movements reveals that the recording was made in vacuum and in an environment with one sixth of gravity, that is, on the Moon.

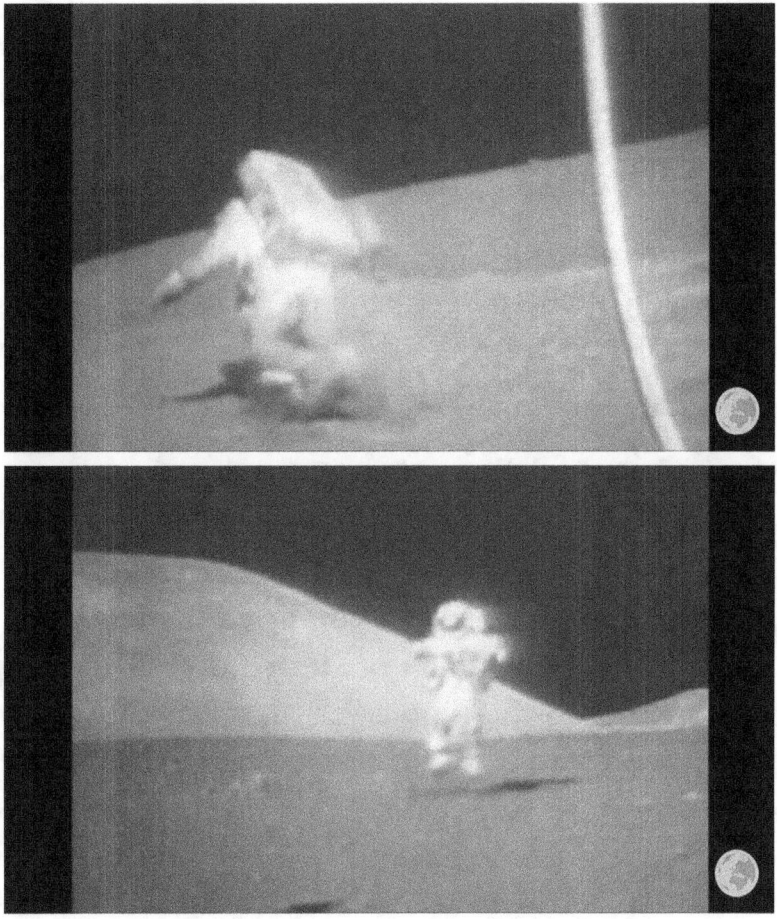

Picture 79 Austronaut´s jump on the Moon

The astronaut's jump on the Moon is seen in the picture above with the legs in straight position. When we are jumping on Earth, we have legs more bent during the jump due to greater force of gravity.

All these recordings are available on YouTube.

Picture 80 Posture of an astronaut in the fall

Today (with newer technology) it is not possible to do such a falsification (editing), and it will never be, except in cartoons or in computer graphic. You should judge it yourself whether computer graphic was used instead of natural recording or not. Would it be possible that computer graphic at the end of the 1960s was more developed than today? For me, this is the most basic evidence that the astronauts were on the Moon.

6.3. The Soviets constantly followed the Apollo journey and confirmed the landing

At that time, the Soviets worked hard to win the race with the United States in the conquest of space, so they carefully monitored each of their development steps. For this purpose (for the journey to the Moon with manned crew), a powerful antenna was built to monitor the journeys of the crew and other telemetric data. Journey to the Moon was confirmed as real. If there had been the slightest doubt that this journey had not occurred, the Soviets would have immediately disproved this event with evidence.

The Soviet Union had self-confidence and courage since it sent Yuri Gagarin as the first man into space in 1961. It also had a lot of knowledge, competitiveness and superiority. They were aware that a flight to the Moon with manned crew was something completely different. At that time of the competition, they did not have any time or resources for this venture, but it was enough for them that they first sent the human being into space. The space race continued peacefully, but later they sent automatic spacecrafts that successfully landed on the Moon. The Americans were aware that they had to secure the safe journey of man to the Moon and take him back safely to Earth. If scientists could not have secured this, they would not have gotten the green light for this project.

The telemetry data obtained by the Soviets during the Apollo flight showed all sections of the journey and the entire timetable of the flight, the flight which was planned with all stages, with individual timetables, and speeds. The plan had to be characteristic only for a flight with manned crew on the Moon and this was also confirmed by the Soviets.

The Soviets even secretly congratulated the Americans, far hidden from the public, although they officially confirmed that the landing really happened.

6.4. It was impossible to pretend genuine concern of people, their cooperation in the Houston Control Center during the Apollo flight

The Space Center in Houston employed 5,000 people, and around 200 people worked directly at the control center. These people were strictly monitoring all the launches from Gemini to the last Apollo. Computers, radars, monitors, etc. were installed on control panels. If flight to the Moon was not real, what would people have been doing there at all and how would they have been pretending and coordinating among each other? This would have taken almost 7 years from the beginning of the development. They had really concerned faces due to errors during launch, not seemingly concerned. These expressions can be evaluated and confirmed by each psychologist that they were genuine.

6.5. Engineering and scientific view on development

A three-thousand tons heavy rocket is freed from Earth's gravity, makes the circle around the Earth carries out experiments of docking of modules over the Earth, and flies to and around the Moon, supposedly leaving out the easier part of the journey (landing on the Moon where there are the same laws as on the Earth with six times less gravity). This is what conspiracy theorists say. So, everyone believes that the rocket did 90% of the hardest work, but then they do not believe that it did 10% of easier work. Those 10% of work are easier from physical point of view, like takeoff from the Earth. Scientists were brave and confident about themselves because of the large number of attempts and tests that had been done.

Nobody would have developed the project if he had not been at least a little convinced that it was workable. Why would someone say "just sit there for 6 years and do nothing". Why would you develop it if at the beginning someone thought that this was impossible? Why would somebody say that despite all successful tests the system would not work? These are reasonable explanations and claims which conspirators do not understand, but they understood the power of provocation.

A constructor who has developed a particular system without or with the help of a computer knows that construction with computer is faster. This does not mean that a constructor can solve more difficult problems with help of computer, but the only difference is that problems can be solved more quickly.

As an engineer, I claim that the ability to build something great, outstanding, is in scientific and engineering minds. Genius and vision are those who "can", and the computer is only helpful in this, and it can be slow or super fast. Super fast computers today help with predictions, simulations to avoid many development stages that people sometimes had to make. With the help of a computer, development can only be faster or done with less number of people. The computer did not create a computer but it was made by a human.

6.6. Computers on Apollo

The operational efficiency of computers was then 10 years ahead of its time. Due to the Apollo program, computer technology has made a lot of progress, thus creating the first digital computer. They were fully operable for all the parameters needed by the spaceship for its journey. The calculations of the movement of bodies in space are based on calculations that had already been known at the beginning of the 20th century. These calculations can be done by a good mathematicians or physicist already with a calculator. Calculators or computers of that time were accurate enough for such a calculation, since it is not important for the accuracy of calculations whether the computer is fast or slow. Years ago, the PCB (Printed Circuit Board) for controlling the launch and ascent of the Saturn 5 rocket (Launch Vehicle Digital Computer, 1969) was shown to the public. By size, this PCB was not bigger than today's with the same capacity, which means that the conspiracy theorists cannot claim that the PCB was not at the level of current ones.

Picture 81 PCB comparison for Gemini and Saturn 5 (The difference is four years of development)

*Picture 82 Comparison of PCBs for Saturn 5 and newer one with the
same capacity (40 years of developmental difference)*

In the picture above, we see how great the progress between the Gemini and Apollo projects was, in the difference of four years. In the bottom, we see the difference between Saturn 5 PCB and the same board of the same capacity. PCB for LVDC for Saturn 5 was so technically perfected that even after forty years, newer PCBs with the same capacity remained the same size. This is the most revealing evidence that in 1969, in the computer field, today's technology for launching the Saturn 5 rocket was already literally being used. Incredible.

The only difference is that at that time the same capable board was much more expensive than today. NASA worked everything with projects, individually, and it could have afforded it because it had lot of money available.

Therefore, I believe that the whole system has done its task more than perfectly. Logical explanation follows that if the conspiracy theorists find today's printed circuit board (PCB) sufficiently powerful and large enough to launch a rocket, the printed circuit developed for Apollo in 1969 must be also sufficiently good for them.

A detailed description of the computer system is described in Chapter 2.6.6.

6.7. The number of people who worked on the Apollo program and the NASA budget

When the Americans sent three men to the Moon in 1969, the NASA budget was worth a good 27 billion of today's dollars. In the years when the mission was being prepared, it was 44 billion (1966). The space agency's budget was shrinking, spreading and turning around over the next few decades. In 2004, Bush declared that NASA would encourage the space program so that it received 19 billion dollars, or 0.66% of the total budget. This budget, however, remained largely unchanged until 2009, with large cuts in 2006. Last year, NASA had just fewer than 18 billion dollars of budget.

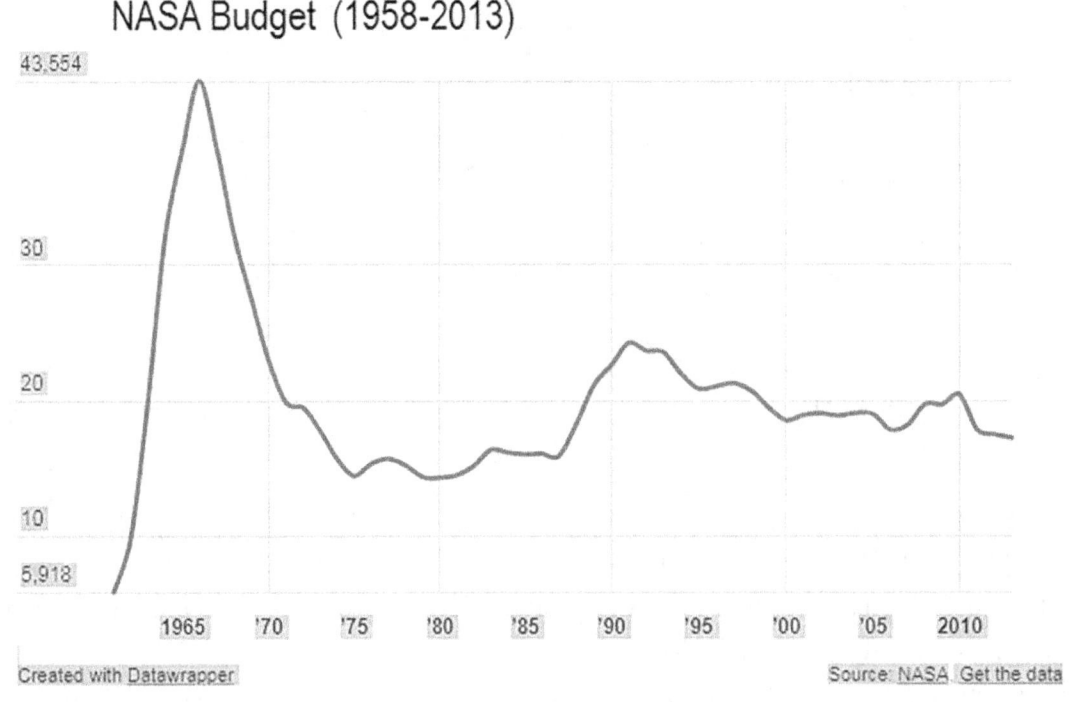

Picture 83 NASA Space budget

For comparison, let us say that the US Army had a budget of 81 billion dollars in 1969, and by 2004, when military operations were already going on in Iraq and Afghanistan, it grew to a staggering 465 billion. In 2013, the US government allocated

640 billion dollars to the US Army which is 35 times more than for its space program. The US Department of Defense allocated 11.5 billion dollars in 2011 for only one project: the F-35 combat aircraft. If we compare the ratio of the 1969 military/space budgets, it is more than clear why the Americans then managed to fly to the Moon.

6.8. Rocks that were brought from the Moon

The picture below represents a glass bead which was located in 382 kilograms of lunar rocks brought by Apollo crews.

Glass beads are created in two key ways: in volcanic activity or in strong meteor strikes that melt and evaporate the rock. In both cases it is necessary for the rocks to cool down and slowly crystallize. On Earth, volcanic glass beads are rapidly decomposed, and in space, glass beads can survive almost untouched. Apollo mission found and brought such beads, which proves that they were really on the Moon.

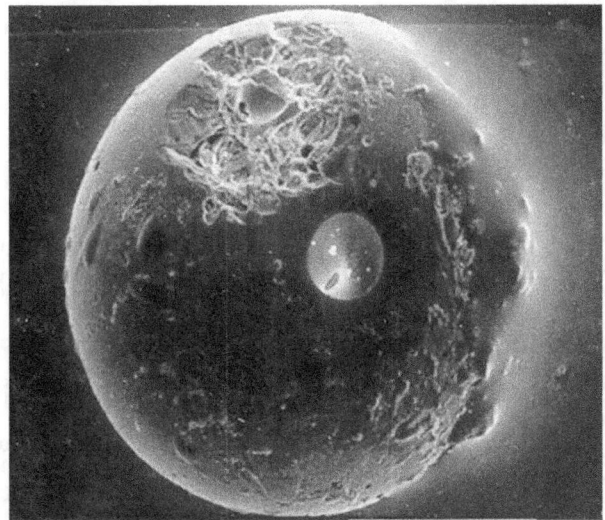

Picture 84 Glass bead without the effect of Earth's erosion

Picture 85 Lunar rock put on display

When the astronauts brought rocks from Apollo missions, these rocks were given to 135 different countries as a sign of goodwill. The rocks have passed all possible analyzes in different countries. Geological researches in these laboratories confirmed that rocks are indeed of lunar origin. No spacecraft itself or an automatic probe could bring rocks in such quantity. In the 1970s, the Soviet Union brought some rocks from the Moon with automatic probes, but they could only bring a third of a kilogram. Both samples from the probe Luna and the Apollo mission match, so they originate from the same place.

6.9. Neal Armstrong's further path

When the first astronauts returned from the mission of the Apollo 11 crew, Neil Armstrong was quite exposed to the public. After a number of interviews, he continued the job of a lecturer at a university. The journey to the Moon changed his personality considerably; as professor he behaved differently. He wanted everything to be precisely planned. He even closed the door by himself after the last students entered the classroom without saying anything to them. He bought a big estate in the countryside so that he could move away from the world and find his peace. Probably, the journey to another planet and the whole stress that he goes through, changes a man greatly. Therefore, the interviews he had on television did not show Neil Armstrong as a man who was relaxed and who had just arrived from the party. Many doubted that his statements were not real and that he was never on the Moon. Neil Armstrong traveled the whole world. He visited almost 54 leaders of the countries of the world who accepted him with state honors. This was an interesting fact that speaks against conspiracy theorists. If anyone had ever questioned this epic journey to the Moon, Armstrong would not have been received by so many presidents of the countries.

6.10. Modern evidence

There are many parts of archive material on all NASA missions in the past and the present. However, I did not want to spend all my time discussing myths about landing on the Moon, without even discussing some of the more recent international missions that prove people really walked on the Moon. The evidence of landing on the Moon with manned crew does not come only from NASA. The Japanese spacecraft Selene orbited the Moon in 2018 and photographed precisely that part of the landscape where a lunar vehicle from the Apollo 15 mission was standing (Picture)

The lunar orbiter took a picture of the Moon in which can be seen on the surface of the Earth's satellite where the lunar modules landed. There are visible footprints of the astronauts, traces of the lunar Rover as well as scientific instruments and footprints of astronauts in the lunar dust of the Apollo 14 expedition. The tools and devices left behind by the Apollo missions have been visible from the lunar orbiter for the first time since then.

Cameras installed on the lunar orbiter, which flew towards the Moon on June 18, 2018, captured five of the six locations of the Apollo missions. The picture below shows images of the Apollo 14 landing site. There are many pictures of the remaining Apollo missions taken by Selene on the internet.

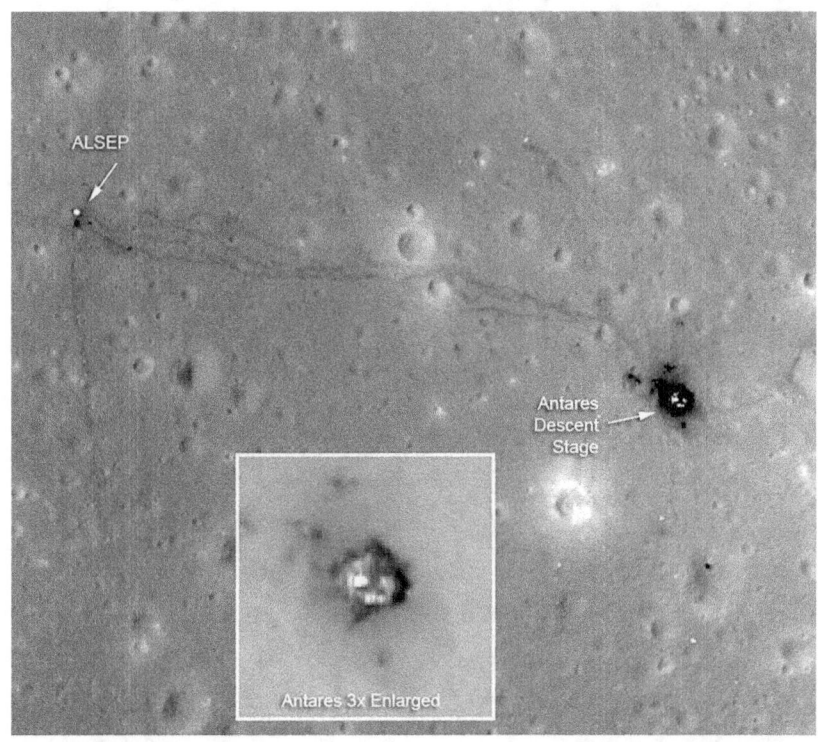

ALSEP

Antares
Descent
Stage

Antares 3x Enlarged

ALSEP equipment

LRV ("moon buggy") tracks

flag

astronaut
footpath

Challenger
descent stage

LRV
final "parking spot"

Picture 86 Apollo 14 landing site

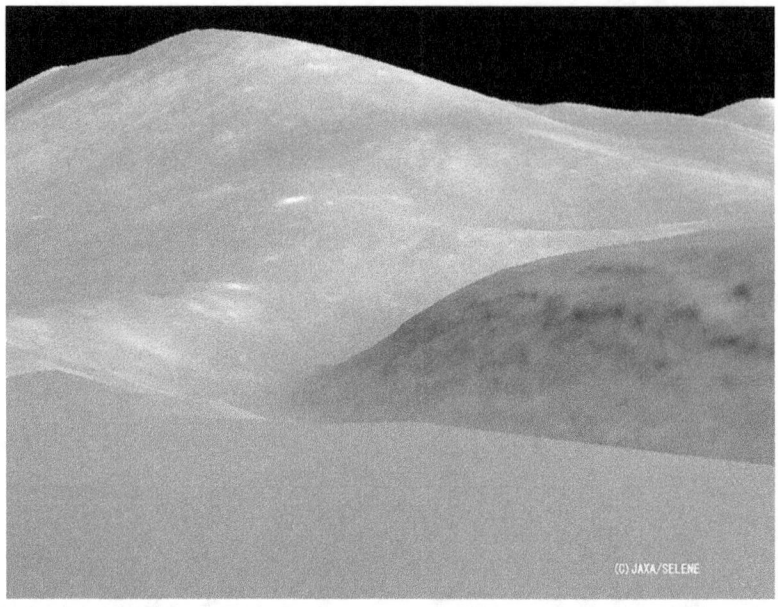

Picture 87 A picture from Apollo 15 and a picture from a Japanese Selene spacecraft photographed in the same plac

The picture below shows the landscape where the Lover Rover was driving. The picture above was created during the flight of the Japanese orbiter Selene in the same place. The landscape is exactly the same, so the lunar Rover stood on the lunar surface.

6.11. Other expert argumentations

1. Film director and manager of Postwar Media SG Collins, has been producing marketing films, television clips and other visual activities since 1981 for various

academic groups that appreciate his complex technology of creation Postwar Media recorded a video detailing how the cassette technology in the 1960s was unable to display a false landing on the Moon. In 1969, there was no film that could record a fake version of the transmission continuously for a time long enough. The live broadcast, which was most viewed in history, lasted about two and a half hours without interruptions.

2. Dr. Davis Grimes, physicist from the Oxford University, has published a theory in which there are three variables

   - The number of people who participated in this project
   - Duration of time since the beginning of the conspiracy
   - The probability of failure of the mission

He calculated that such a mission, given the facts, was impossible to be falsified. With nearly 400,000 people who took part in the project, duration time of the discovery of conspiracy would only last eight months. Now it has been 50 years since then, and none of the participant has discovered anything. If this was a hoax, serious evidence of the conspiracy would have already been presented.

3. In 1969 the following scientific instruments were available for exploring the Moon

   - Gravimeter
   - Ultraviolet spectrometer
   - Infrared radiometer
   - Radar
   - A specially developed navigation device for measuring the position of the LVR to LM
   - Device for measuring the heat flux
   - Solar wind analyzer
   - Seismograph, TV camera controlled from the Earth
   - Magnetometer
   - Device for measuring cosmic radiation
   - Device for measuring seismic activity
   - Laser refractor
   - Device for measuring the heat flux

All these scientific devices are topmost and suitable for such researches and measurements. Nobody said that at that time, these devices did not exist.

4. Numerous panoramic images from the Moon

Astronauts recorded a multitude of different panoramic images on the Moon; landscapes and rocks of various shapes. How could the scenery have been faked for such a large number of shapes of landscapes and rocks? Hundreds of people, who could have told something about this theme, should have participated in these sceneries.

6.12. Where could such a large rocket take off to?

If we look at the picture below, we can see (this is believed by the whole world) a rocket as big as a skyscraper, smoke that twists miles away and the unimaginable rumble is heard far away. Where could such a rocket go? For orbiter into Earth's orbit, a five times smaller rocket would have been big enough. It wouldn't have been possible that they hid it into Earth's orbit. On Earth, in the Atlantic Ocean, only the first stage of the rocket was found, and no one ever found a second or third stage or service module, since these parts actually remained in the space outside the Earth's orbit. Were these parts hidden in the space? Of course not, for what reason? They were namely so close to the Moon that they just only had to land on its surface.

Picture 88 Launch of the Saturn 5 rocket with Apollo 11 spacecraft

Some data on the size of the rocket:

- It is 120 meters high, 20 meters higher than the Statue of Liberty.
- Filled with fuel is thirteen times heavier than the Statue of Liberty.
- If it had exploded on the pad, more than 2.7 million kg of fuel would have exploded with a power of 610 tons of Trinitrotoulene.
- It took 96 railway tanks to fill the Saturn 5 tank.
- The tank of the first stage of the rocket is so large that there can stand parallel three giant trucks in It.
- Engines of the first stage produce as much power as all the waters of North America if all of them were installed through turbines at the same time.
- More than a million parts of Saturn 5 had to work harmoniously.
- During the takeoff, a giant was poured with 220,000 liters of water every second.
- Five Saturn engines collected thrust power of 92,000 locomotives or 500 000 cars during the takeoff.

Who would build a device with such impressive properties just for deception? What arguments could the conspiracy theorists give for this information? This takeoff actually happened, and conspiracy theorists and millions of people, who watched the takeoff, know that.

6.13. Summary of evidence

The evidence I provided was not easy to write at once, the ideas have been emerging for more than a year. Since I am convinced of a real landing on the Moon, I have been thinking that there must be something to prove the case. Because if something really happened, it must leave the evidence that we have before our eyes, but we do not see them immediately. The most convincing evidence is the easiest, but the most difficult to imagine. In simplicity there is genius. The scientists who enabled this journey were simple and at the same time genius people.

7

Political and economic point of view

7.1. Why one has not set foot on the Moon since 1972

THE FIRST FLIGHT to the Moon in 1969 was a matter of political prestige. The competition between the two superpowers came after the Second World War, after the defeat of Nazism. The display of power in the military sense or the planned attack of one superpower on another would have made no sense. That's why a prestigious duel on nuclear weapons occurred, or who could make a stronger atom bomb. These bombs, or their destructive and radioactive strength, exceeded the necessary power to destroy the entire world for several times.

Somehow they had to choose an area where they could display technological superiority without the danger of destroying the world. This was possible to achieve with the development of space technology, initiated by the Soviet Union. Without these events, we would be still probably not talking about landing of a man on the Moon today.

Of course, this competition did not serve to have direct economic benefits. These, however, later turned out to be the impetus for the development that would serve the future of mankind, but the most America itself.

America still has political and economic benefits from it. Perhaps pure politicians and economists are not even aware of this. However, the other countries seem to just leave the USA aside, that they should be the first to come up with new technological inventions, because it is the belief that only they can do this and that they have huge amounts of money and unlimited human capacities. This is a great advantage nowadays. For me personally, the flight to the Moon is still a continuing economic benefit.

The last flight of the human crew to the Moon took place in December 1972 with Apollo 17. They planned flights up to Apollo 20, but they probably did not realize it because of their low interest. The flight of the Apollo 11 was something extraordinary, but that flights were then repeated every half-year, this already borders on science fiction. Nowadays, barely one flight would get a political permission.

The United States achieved its goal ambitiously, and after that, in 1973, the Saturn 5 rocket was used for the last time to launch the Skylab space station into the Earth's circumference.

The Soviet Union congratulated the United States, acknowledging its superiority, followed by the decreased political and public interest in conquering the space. Other demands have come to the fore, oil crisis has taken place, and interest in oilfields has increased, especially in the Arab world. The competition moved to the military field. Nuclear weapons, ballistic missiles, etc. were developing. I, personally, and probably also mankind, regret that some rapid technological advancement in the conquest of space has been abandoned. In the early 1980s, the Space Shuttle project came to life, but it did not even reach such a prestige as the flight to the Moon.

In the 1970s, the projects Viking and Pionner continued after the Apollo, which at least satisfied the scientific public a little. Nowadays, projects continue with automatic research devices on Mars, and fly-bys on other planets.

After the Apollo program had ended, ordinary people and scientists were so enthusiastic that they predicted that a man would have stepped on Mars in 2000. None of this happened, this was a negative progress. It was no longer in the interest of the leaders of countries that were able to carry out space conquests financially and technically. It became much easier that they preferred to be dudes in their own or in neighbour's garden. This is also due to the political and economic system that has even more expressed the functioning of the market, supply and demand, and also prestige in the financial markets. A kind of economic and banking elitism has come to the forefront, which, of course, works only for its own benefit and not for the benefit of mankind. The elite have such power that politicians just only follow their needs.

In recent times, competition has been felt again, who would be going where and whether they would send a man back to the Moon, although the main future goal is to fly to Mars with a human crew. For our nearest natural satellite there are several players, the USA, Russia, China....

7.2. Further explorations of the Moon

Explorations are becoming more and more expensive and money is also more and more available, we have just to decide where would money go and for what cause. The value of the journey to the Moon is about 30 x the value of the flight into the Earth's orbit. Journey to the Moon by setting up a base and returning from the Moon with a human crew today would cost over one hundred trillion dollars (one hundred billion dollars_EU). The technology became cheaper but the space did not.

The space game is slowly developing, not with such an impetus as during the Cold War though, but the tension between China, the United States and Russia is slowly and verified growing. According to the facts, the Chinese space program has a rapid impetus and employs as many people as they work at an international space station project. (16 countries altogether) With the arrival of China, everything has changed. They are not burdened with the Cold War space competition which was just a prestigious competition. With them, each step is greater than in the case of the Russians or the USA, since they are not burdened with history, and the burden if it is worth investing in space or not. Obviously, during the regression of the previous superpowers, they have accumulated enough knowledge to be able to compete with them successfully. Time will tell! For now, the Chinese are still lagging behind, but according to what they have been able to do in less than 10 years, the next 10 years will be particularly interesting.

In case of Taikonaut's landing on the Moon, the United States would feel threatened, Russia defeated, and other countries would fall behind the red dragon. The USA must therefore gain some kind of "advantage" over the Chinese.

The statement that the Americans brought only bare rocks from the Moon is not true, because Chinese are also interested in these rocks containing the He3 element. That is why the scientist (geologist) who has ever walked on the Moon, Carel Schmitt, is the main proponent of the return of a man to the Moon.

Russia would also feel bad, because their knowledge from China's point of view would be completely overcome.

In short, it's not a matter of money, but a matter of human way of thinking.

Lately, we have heard that the Chinese have discovered large reserves of various ores, gas, etc. on the Moon, so they are rapidly preparing the expeditions to the Moon. But it is interesting because it has not been heard about any ore resources until now; only desolate rocks that the Americans brought to Earth. Behind all this, the non-radioactive isotope He3 is hidden, which would solve the problems of nuclear fusion and thus the energy problems on Earth.

The Chinese are preparing themselves for the Moon, also because they want it. They do not plan flights according to presidential elections, but according to the timetable. The Chang'e 4 automatic ship landed on the back side of the Moon. A special satellite provides the radio connection, because the radio connection does not work directly with the back side of the Moon. At present, the only reason for establishing a more permanent base on the Moon is therefore He3. But this is in solid form of ice or minerals. By acquiring this isotope and by establishing a base, it would also allow deeper space travels with significantly lower costs. With Chang'e 5, the Chinese will bring some of these minerals to Earth. There is no need to rush here, since the first thing that needs to be done is to develop a fusion-based reactor. This may take several years. However, this date also matches the dates of the first flights to the Moon. Countries in the ITER project are also countries that are directly or indirectly linked to the He3 competition. Once the matter is settled on Earth and the fusion reactor is improved, the return journey to the Moon will be even more topical.

Today, there are no special political reasons for the flights to the Moon and there is not also any huge capital required by the development of these flights, the countries do not want to invest anyhow. Private capital requires immediate profits, and for the moment, it does not invest much in these programs, as it is not clear when the profit would be available.

Russia also plans to send expedition with human crew to the moon by 2025. They are supposed to set up a permanent base. It is completely realistic to expect that Russians could send people to the Moon by 2025. We see that in ten years since a man has flown into space, he has already walked on the Moon, and then it is also realistic to expect that the matter can be accelerated and the steps can be taken 3 to 5 years earlier.

All the predictions of these countries (China, USA, Russia, even Japan and the EU) are crossing during 2020 and 2025!

Anyway, time will tell its own tale, but we will again go to the Moon with a reason, and this reason will be He3, and a scientific research base with a large telescope on the dark side of the Moon. On that side, radio telescopes will be installed for "eavesdropping to the space", as there will be no disturbances coming from the Earth.

Whoever will step next on the Moon will be the second...because America has already been, but whoever will control the Moon will control the world in the next 100 years, of course. Why are they even wondering if they would return to the Moon and whether this journey is worth it? Yes, videos would be much more interesting than watching reality shows. In the development, there should be no immediate primacy for immediate economic benefit. If you›re not curious, and if you do not have empathy for new, unknown, you're not a worthy person. With economic logic, we are getting more and more poisoned. It does not pay off anymore if it does not remain a substantial part of the profit.

Let us release exploratory outlook, move ahead, and realize what we people dream and what we want. We do not always say that the most important thing is work, job and money... We do everything much more than we need at all. Anyway, we are too busy and diligent, and let us be overcome by dreams once again.

Lately, the atmosphere has been completely changed by the visionary, engineer and businessman Elon Musk, who set the goal of bringing a man to Mars. What is the ultimate goal, and what the goal of mankind will be, is described in Chapter 9 (Future).

8

Moral of the story and conclusion

THE MOST MENTIONED reproach that in 1969 they could not fly to the Moon was because of the underperformance of computers at that time. I described the principle of the operation of the Apollo computer which was good enough for this task in Chapter 6. Doubts about similar achievements have already occurred in times of dark history for centuries. This is a cross-incompatibility, so that one doubts quite a lot and thinks too little.

The causes can also be found in the too commercialized current generation. It is believed that it is superior and the first one in which real changes will really start and that they are the ones who will change the world. The commercial industry and general greed are the cause that separated this generation from its essence and its foundations. The previous generations, and especially the generation X, was one of the most advanced generations of history in its impetuous rise. This impetuosity resulted in large development leaps of the so called vertical development compared to the previous twenty years.

Today's development is based primarily on computer science, the development of many applications that shorten development and production processes. I call this a horizontal development, which has not yet given such a breakthrough for a person to feel this as a greater own happiness.

**Vertical development (part-time employment, healthier
life, faster journeys, journeys to other worlds)**

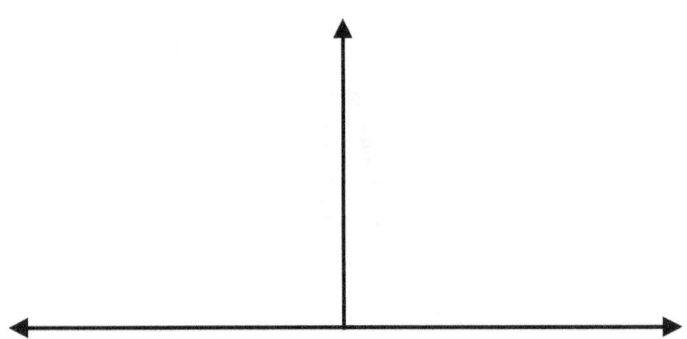

Horizontal development (faster computers, more applications in a short time)

Horizontal development does not directly affect vertical development. A man is the one who determines the vertical development, what he wants and does not want. In 1969 they wanted to fly to the Moon, landed there and returned to Earth. And they did. We see the direction and the way of today's development as a problem, because the development itself does not follow ethical and moral progress. It is taking place on a horizontal line, which can go a long way. However, this development is little successful only on the vertical line.

The introduction of 5G technology is a typical example of horizontal progress that threatens us and which could be a step backward in ethical and moral terms.

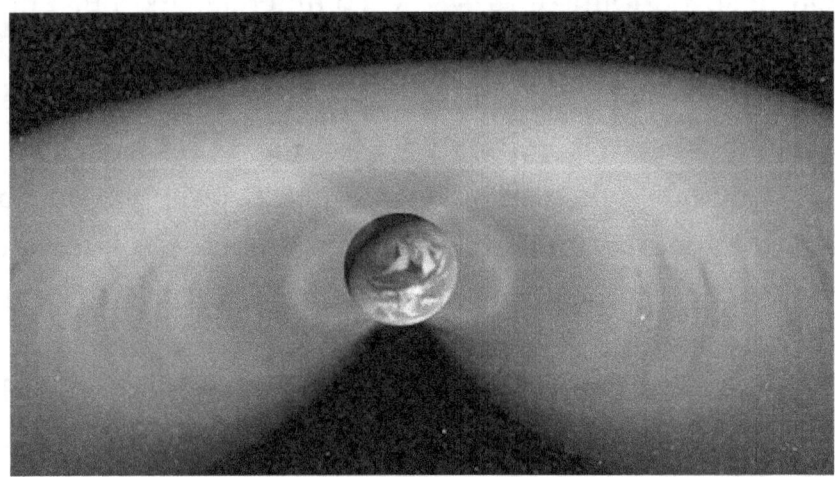

Picture 89 Saturation of the world with 5G microwaves

The world will be more than ever intertwined with high frequency microwaves, 90% of which comes through the skin. The impact on human health and human dependence on 5G technology has not yet been investigated.

If you really want to make progress, you have to have a healthy and happy person beside your goal. Today's generation has received the basics of computing from previous generations, and the latter have put them on the highway in the direction of progress so that they can now just drive forward. I hope that progress will be pushed forward in the right direction and we will be grateful if all previous achievements are properly upgraded. I am afraid that now they are only comfortingly considering what could be done with bits, ones and zeros, and that because of this, people are even more burdened with information with less and less free time. But do you expect that we will be developed only by sitting in front of computer for hours, where we type hours and hours on keyboard?

It will be easier for us when we will have more free time because of technology and maybe enjoy watching transmission of humans walking across Mars.

For now, the technology has made so much progress that it is possible to order a pizza home from a smart phone, or to find out where your postal item is currently located... 5G...

The computer was been developed for personal use even more rapidly at the end of 70s. Office work has become easier, writing has been faster, and a big step has been made in saving a multitude of documents. As soon as you wanted to change the old document, you could do it right away and save it again. In the entertainment area, computer games appeared which in the fictional world gave them a sense of control and superiority. The possibilities were practically infinite. The sense of power and progress was unlimited. Everything that was not connected to the computer was reactionary. As soon as somebody showed annoyance and lack of interest or fear in front of the computer, he was labelled as backward. Digitalization has started to occupy various areas in all social segments. The emergence of the Internet has only accelerated all this, accelerated the sense of superiority over all of the previous things. It is believed that everything that has happened before has turned out to be not good enough, too slow, or obsolete. The doubters are wondering how they could even fly to our natural satellite if they did not know the computer skills well? (Of course they did). For people who do not understand science this seems impossible, but, of course, it is possible because the necessary mathematics for flying between celestial bodies was created by mathematician Tsiolkovsky, Goddard and many others at the beginning of the 20th century. These formulas can also be solved with a calculator.

Computers were not that bad even for present situation in Apollo. In one of the aforementioned evidence, chapter 6 describes the motherboard for regulation of launch and speed for the Saturn 5 rocket. All this was feasible, since thousands of people worked for this project alone at that time. Today, such manner would be too expensive, but back then, with perseverance, the goals were achieved. As a writer, however, I wonder how it is possible that such a gap was left between today's understanding of the then development.

The media does not devote much attention to one of the most epic journeys in history anymore. What is only considered is a rapid rise of financial elite, fashion industry, celebrity and popularity, and also oil, military industry and hedonism. This is certainly one setback from a moral and ethical point of view.

Everything has become so connected and global that we cannot imagine life without a computer, so this generation has a great impact on public opinion. There is also some envy about the success of Apollo, and this has become the perfect formula for doubt that has arisen and continued to grow. Of course, this doubt has nothing to do with wisdom, nor does it have a computer that has become the mental key to doubting the flight to the Moon. The other thing is that it is necessary to ask ourselves if everyone can judge whether each modern educated man is capable of thinking and evaluating what is right. Society has a crisis in this respect. The disconnectedness and lack of interest among the generations is so strong that today's generation does not know anything about what and how the development was carried out by the generation that landed on the Moon. We must be more and more aware that in 1969 occurred a remarkable event in the history of mankind. I do not know with what achievements this can be compared. Civilization has shown what it can do when it combines the knowledge and the will to achieve one goal. We have to be grateful to these people and to their enthusiasm. The impeccability of the Apollo program, despite the disasters, was largely by merit of devoted people who had planned it; they had a huge amount of knowledge, from whom many people can still learn today. This ability reminds us of the incredible achievements of small beings such as ants, as all of them as a group operate as one system, and the result is the incredible structures which they are able to build. The successes of such projects depend primarily on people. I would say that computing can contribute somewhere up to 10% to the success of the entire project. Digitalization, however, contributes to the fact that some development problems can be solved more quickly, which is again not entirely true, as it has not been proven that projects are sooner operable and more reliable in recent times. When I started my job, I designed some mechanical systems without a computer, and in recent times, I have been modelling systems with 3D CAD. I do my job faster now, but my computer does not allow me in any case to be able to solve difficult problems for that reason.

Despite the above mentioned, many things are changing. Just to be clear, in spite of everything, the most useful things that the world needs are being done now. But we are still at a crossroads, which has two directions: Better life for all people or increasing burden. After landing on the Moon in 1969, we felt that the world was heading towards a better life for all. How it is today the reader should decide on his own. I believe it will be better, since a man is always able to make a right choice at last minute.

AT THE ENF OF THE MORAL OF THE STORY

The future has already happened; we are now waiting for a new future.

Our job is to continue where they ended, so we can move forward.

What now? We stepped on the Moon 50 years ago, some do not even believe that we really were, and now it's even more unclear how to move forward. The world has not been so irresponsible, as it is now.

Why they cannot get to the Moon after 50 years with better technology? What better, they do not have it!

What kind of technology was available that they landed on the Moon and returned to Earth? Good enough that they did this.

Better computers do not mean better technology in rocket engineering but only a lot of exceptional engineers and scientists were needed.

Despite today's technological "superiority" and knowledge, we do not know much about the Apollo program, which was the most complex project in the history of mankind.!?

Which today›s president in the world can pride himself on an act like President Kennedy, who gathered practically all the people of the world in front of screens?

My daughter helped me to write the book and asked: "Dad, is it really true that they were last time on the Moon in 1972? How strange is that, she replied."

The epilogue was done by one American scientist in a broadcast on Elon Musk: I was 17 years old when I was watching the landing on the Moon, and if someone

had told me at that time that we would still not have come to Mars in 2000, I would have said say that he was crazy.

A complete change in technology of such dimension can change every hundred years. Employees in NASA in the sixties were able to show the way to all future generations. When will we outshine them?

Compare the possible saturation of the world with 5G microwaves, or the extraordinary journey of people to the Moon.

Even today, some things can be taught by ancient philosophers, and their wisdom is gaining more and more importance.

Galileo Galilei said for Earth: AND YET IT REVOLVES
I say for brave astronauts: AND YET THEY WERE

9

Future

WE CAME TO the Moon 50 years ago, and at that time it was almost a fact in the scientific circles that we would go to Mars at the latest in 2000. We would be strange if we predicted otherwise. As one scientist said; the development of rocket technology has gone back, which is something incredible. Now that we have realized this, we slowly obtain the money and the power once again to replace the lost.

Elon Musk, perhaps NASA, maybe joint international mission, are preparing for this step, the departure of a man to Mars. The problem with Mars is that it is 300x further than Moon, and only barely every two years it is a good time to travel there because it has to be in a favorable position relative to the Earth. The problem is also that enormous amounts of fuel are needed for the flight to Mars and back. Mars travels around the Sun with 30 km per second, while the Moon travels around the Earth with 1 km per second. So the problem is not only a larger distance, but rather that the rocket needs a lot of energy to catch Mars' speed so that it can be orbited into its orbit. Traveling there and back lasts almost two years, while taking care of food, water, oxygen and an appropriate energy source.

Regardless of everything, in the future people will become interplanetary civilization because the survival on Earth will not be possible in the long run. We could face wars, atmospheric poisoning, overpopulation, water and food shortages or drastic climate changes. Many people are asking themselves why this is necessary. I know that human is not destined only to Earth, now and forever, so it is necessary to begin planning interplanetary missions. Timeliness in finding new homes can ensure the survival of our civilization.

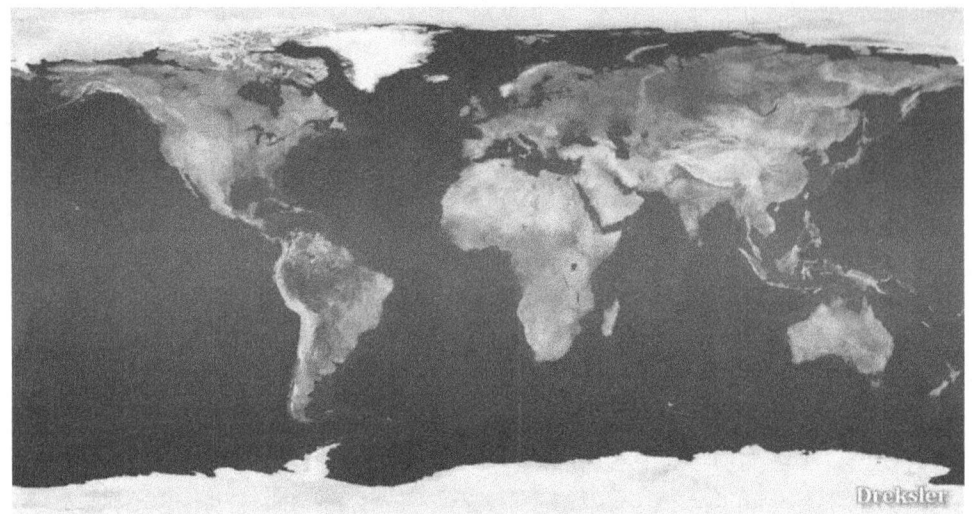

Picture 90 The Earth today

Of course, Mars is not favorable for immediate settlement of earthlings, but it needs to be gradually adapted. Previous research has shown that the red planet is favorable for terraforming. Terraforming means the adaptation of a foreign environment to life as we know it on Earth. There is enough water on Mars; it is under surface and mostly in the form of ice. Firstly, it is necessary to raise the temperature so that the water begins to change the state of matter from solid to liquid and then to gaseous as evaporation. Evaporation means the rise of oxygen in the atmosphere, the rise of air pressure, and consequently the formation of clouds from which rain can be formed. When the atmosphere becomes denser due to evaporation, it retains more heat and the process is repeated, the temperature of the atmosphere gets higher. To raise the temperature, black soil or coal dust could be scattered on the equator in order the surface to absorb more solar heat.

Oxygen can also be obtained from CO_2, which is almost 98% in the Mars' atmosphere, as CO_2 also consists of oxygen. We need to find such plants that would grow in the Mars' soil in order to produce food. Mars' soil is already suitable for certain terrestrial agro-cultures and contains some nutrients, as confirmed by scientists.

In 800 million years, due to the process of the fusion reactor in the core of the Sun, the Sun will be warmer by 10%. On Earth, the average temperature will be 50 degrees centigrade, which means stronger evaporation of water and denser and more humid atmosphere, which only increases the sensation of heat. Life will be unbearable on

Earth, and Mars will find itself in the Goldilocks zone (fn. the habitable zone), and our civilization will have to move to Mars.

Picture 91 The desert on Earth today

Picture 92 The desert on Earth in a billion years

The settlement of Mars in the future is inevitable. With well-planned procedures, terraforming can last from 100 to 200 years, which is enough to adapt Mars to living conditions comparable to those on Earth. There is more than enough time. During this time, sufficient energy sources can be found for the permanent settlement of people on Mars. In the beginning, the solar radiation will be a major problem, since Mars' atmosphere does not currently retain radiation from the space.

Before settling, it is necessary to find a suitable source of energy that we must first obtain on Earth to produce it later on spaceships, and ultimately on Mars itself. What did the flights to the Moon with manned crew tell us? That everything can be done with careful planning and will.

Things will become more and more complex; we will use powerful computers that will be indispensable in the future because they will be able to predict relationships with the simulations, which we do not know yet.

In 4 to 5 billion years, the Sun will become a Red Giant; the internal pressure will rise greatly, so the Sun will blow up to the Earth's orbit. The balance will be restored when helium begins to flow into heavier elements. At that time, the Sun will most likely swallow Earth, and there will be scorching heat on Mars.

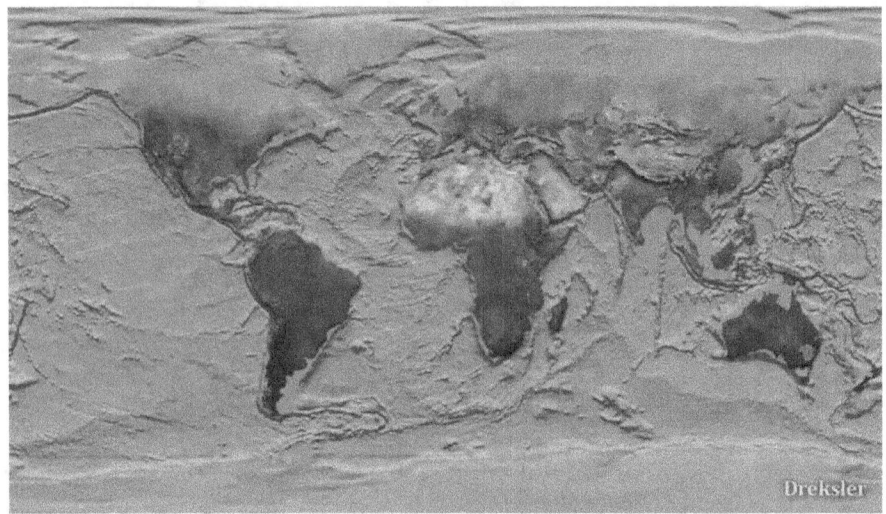

Picture 93 The oceans will evaporate due to the heat

Picture 94 Red Giant and conditions on Mars

Mars will also become too hot and civilization will have to migrate even further from the Sun to the life-sustaining zone.

The Jupiter's moon Europe would be very suitable, which now consists of a frozen ocean of water. In the case of the Sun as a Red Giant, oceans of running water would be created on the Moon Europe.

Picture 95 Red Giant as seen from one of Jupiter's moon

Even more suitable for survival would be Saturn's moon Titan where it is now -179 degrees C. Then, the temperatures there would be the same as they are now on Earth.

Picture 96 Conditions on Titan

Picture 97 Conditions on Titan

In 10 billion years, the Sun will cool down when all the energy available is consumed. Temperatures on settled planets will fall again, and this decline may last a billion years. During the cooling of the Sun, we could resettle Mars if everything was not going to be destroyed by the heat of the Sun until then. Otherwise, when the Sun is completely extinguished, maybe we will find a planet in the solar system, or a satellite of one of the planets that could be settled below the surface if there were suitable temperatures due to the hot core.

From then on, mankind will have to find a suitable drive for interstellar flights. We already know of at least 200 planets in our vicinity to a distance of 200 light years, of which at least 20% are similar to Earth, and quite a few are in the Goldilocks zone where there could be running water. These planets must also have a magnetic field that would protect the planet against dangerous particles of radiation from the parent star.

The flight to the Moon was the first step of a man to explore a foreign celestial body with manned crew. It has become a pioneer for all further journeys. From these first steps in 1969, we learned how careful preparations were and how no procedure should be left to chance. All procedures were planned and completed. Let us be satisfied and happy that someone has already paved the way for us, we just have to follow their enthusiasm and their dream. Humanity is on the path of solution and existence also because of them.

If mankind will not travel in space and look for new homes, it will be extinct because the Earth will become an unfriendly world.

And here is something else about the end of the universe and my personal opinion contrary to other theories. If universe spreads, it will stretch until it decays into atoms, to the simplest hydrogen atoms. Let us hope this will again sow seed for the emergence of a new universe. Hydrogen atoms as the building blocks of the universe could begin to merge again and produce heavier elements if there was enough gravity, of course. The stars, the planets would be born again... Beautiful, maybe real dream...

www.ingramcontent.com/pod-product-compliance
Lightning Source LLC
Chambersburg PA
CBHW080958120626
46546CB00010B/2947